NEBS MANAGEMENT DEVELOPMENT SUPER SERIES

THIRD EDITION
Managing Activities

Preventing Accidents

Published for
❦ NEBS Management *by*

Pergamon Open Learning

Pergamon Open Learning
An imprint of Butterworth-Heinemann
Linacre House, Jordan Hill, Oxford OX2 8DP
A division of Reed Educational and Professional Publishing Ltd

 A member of the Reed Elsevier plc group

OXFORD BOSTON JOHANNESBURG
MELBOURNE NEW DELHI SINGAPORE

First published 1986
Second edition 1991
Third edition 1997

© NEBS Management 1986, 1991, 1997

All rights reserved. No part of this publication may be reproduced in any material form (including photocopying or storing in any medium by electronic means and whether or not transiently or incidentally to some other use of this publication) without the written permission of the copyright holder except in accordance with the provisions of the Copyright, Designs and Patents Act 1988 or under the terms of a licence issued by the Copyright Licensing Agency Ltd, 90 Tottenham Court Road, London, England W1P 9HE. Applications for the copyright holder's written permission to reproduce any part of this publication should be addressed to the publishers

British Library Cataloguing in Publication Data
A catalogue record for this book is available from the British Library

ISBN 0 7506 3706 4

Whilst every effort has been made to contact copyright holders, the author would like to hear from anyone whose copyright has unwittingly been infringed.

The views expressed in this work are those of the authors and do not necessarily reflect those of the National Examining Board for Supervision and Management or of the publisher.

NEBS Management Project Manager: Diana Thomas
Author: Joe Johnson
Editor: Fiona Carey
Series Editor: Diana Thomas
Based on previous material by: Joe Johnson
Composition by Genesis Typesetting, Rochester, Kent
Printed and bound in Great Britain

Contents

Workbook introduction v
   1   NEBS Management Super Series 3 study links   v
   2   S/NVQ links   vi
   3   Workbook objectives   vi
   4   Activity planner   vii

Session A Accidents and their causes   1
   1   Introduction   1
   2   Definition of an accident   2
   3   What kind of accidents?   4
   4   What causes accidents?   6
   5   Policies for safety   12
   6   The team leader's role   14
   7   **Summary**   16

Session B The management of safety   17
   1   Introduction   17
   2   The cost of accidents   17
   3   Management strategies for safety   19
   4   Accident prevention and the law   23
   5   Risk assessment   26
   6   People with a special role to play   34
   7   **Summary**   39

Session C Practical accident prevention   41
   1   Introduction   41
   2   Machinery safety   41
   3   Preventing falls   45
   4   Electrical hazards   49
   5   Maintenance work   52
   6   Manual handling   53
   7   Fire hazards   59
   8   Protective equipment   60
   9   Day-to-day tasks   63
  10   **Summary**   67

Session D Coping with accidents   71
   1   Introduction   71
   2   Dealing with accidents and abnormal occurrences   71
   3   Reporting accidents   78
   4   Investigating an accident   79
   5   **Summary**   83

Performance checks — 85
   1 Quick quiz — 85
   2 Workbook assessment — 87
   3 Work-based assignment — 88

Reflect and review — 89
   1 Reflect and review — 89
   2 Action plan — 92
   3 Extensions — 94
   4 Answers to self-assessment questions — 96
   5 Answers to the quick quiz — 99
   6 Certificate — 100

Workbook introduction

1 NEBS Management Super Series 3 study links

Here are the workbook titles in each module which link with *Preventing Accidents*, should you wish to extend your study to other Super Series workbooks. There is a brief description of each workbook in the User Guide.

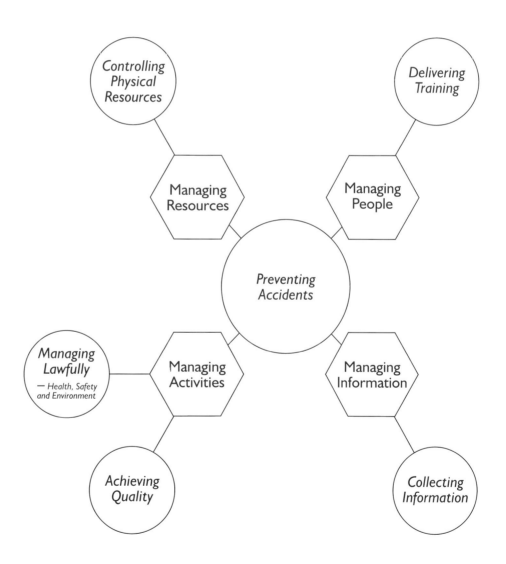

Workbook introduction

2 S/NVQ links

This workbook relates to the following elements:

    A1.2   Maintain healthy, safe and productive working conditions
    D1.1   Gather required information
    D1.2   Inform and advise others

It is designed to help you to demonstrate the following Personal Competences:

- building teams;
- focusing on results;
- thinking and taking decisions;
- striving for excellence.

3 Workbook objectives

'Work can make a positive or a negative contribution to a person's health. When people are exposed to danger (for example, in the form of exposure to chemicals, certain repetitive tasks or a risk of falling) physical and mental health may suffer. In the absence of danger, when people are interested and involved in their work, satisfaction and enjoyment are increased, and improvements in health and wellbeing can result.'

Health and Safety Executive booklet
Successful Health and Safety Management.[1]

You don't want to expose your team members to danger. Accidents are costly (measured in both human and financial terms), disruptive, and morale-destroying. If you've ever been involved in a serious accident yourself, or seen someone get hurt in one, you won't have forgotten it in a hurry.

As a manager, you have a responsibility to find ways of preventing accidents, and of minimizing the risks from hazards at your place of work. The philosophy of accident prevention is, in essence, simple: identify the hazards, and then put all necessary measures in place for eradicating them, or at the least, protecting people from them. As we will discuss, most accidents at work are the result of a failure to put this philosophy into practice in an adequate manner. In other words, accidents usually occur because the health and safety management system breaks down. To put it even more plainly: the majority of accidents could be prevented, if safety were better managed within the organization.

---

[1] HSE Books, 1991, page 6.

Workbook introduction

This workbook has four sessions. In Session A, we'll take an overview, and define what we mean by 'accident', 'risk', 'hazard', and so on. After looking at a number of descriptions of accidents, we will try to identify some of the causes, and the means of preventing similar accidents.

Session B examines safety from the point of view of management: costs; system strategies; legal obligations; risk assessment; people with a special role.

In Session C, we get down to practical accident prevention: analysing different types of accident, and identifying hazards common to many workplaces.

Session D is entitled 'Coping with accidents'. It looks at the activities that must take place once an accident has occurred: emergency procedures, reporting, and investigation.

3.1 Objectives

When you have completed this workbook you will be better able to:

- play your part in implementing and maintaining safe systems of work;
- identify hazards in your workplace, and take effective precautions against them;
- take part in risk assessment;
- identify some important points of health and safety law;
- cope with, report on and investigate accidents at work.

4 Activity planner

The following Activities require some planning, so you may want to look at these now.

For Activity 6 you will need to obtain a copy of your organization's health and safety policy statement.

Activity 11 asks you to think about the status of your team's training on health and safety. It would be useful to identify beforehand what training the team has had.

For Activity 15, you are expected to give examples of identified hazards at your place of work, the results of the last risk assessment of these hazards, and to describe any further actions you plan to take with regard to them.

In Activity 28 you are asked to use a checklist to assess a particular manual handling operation. You should try to obtain a copy of the HSE booklet *Manual Handling – Guidance on Regulations*.

Workbook introduction

Activity 34 requires you to undertake a thorough review of the accident prevention measures currently in place in your work area, in respect of a chosen topic.

Activity 35 asks you to note two or three examples of accidents that you know have occurred in the past, and also to try to think of situations where an accident came close to occurring.

Portfolio of evidence

Some or all of these Activities may provide the basis of evidence for your S/NVQ portfolio. All Portfolio Activities and the Work-based assignment are signposted with this icon.

The icon states the elements to which the Portfolio Activities and Work-based assignment relate.

The Work-based assignment requires you to carry out a risk assessment. This task is designed to help you meet element A1.2 'Maintain healthy, safe and productive working conditions'; it may also be useful in contributing to elements D1.1 and D1.2 of the MCI Management Standards: 'Gather required information', and 'Inform and advise others'. You may want to prepare for the assignment in advance.

Session A Accidents and their causes

1 Introduction

> On average, **ten** people are killed at work **every week** in this country.

Accidents happen everywhere: in the home, on the road, in sporting events; the world has never been a completely safe place. But we expect a place of work to be a controlled environment, where everyone has a defined role to play, and which operates according to a set of rules. Here, if anywhere, the actions of people, and events, should be regulated and well-organized. And yet, alas, accidents at work continue to cost hundreds of lives each year, and thousands of injuries.

This fact can't only be explained by lack of resources: tough bosses thinking only of profits, rather than the safety and welfare of their employees. In many industries, organizations invest enormous sums in making workplaces safe and healthy. Go into any large chemical works, for example, and see the systems, the precautions, the continuous training programmes; you will hear and read the words 'safe' and 'safety' many times during a day. But accidents **still** happen, even in the best-run workplaces. Many occur because people, for some reason, neglect to take simple precautions; others come about as a result of poor supervision. Whatever the cause, it must concern management, for it is managers at all levels who have the task of implementing systems of safety, and of ensuring they are maintained.

We begin this workbook by defining what we mean by an accident. Then we'll list a number of 'classic' accidents, and look at some examples.

Next, we'll start to examine causes, in order to help us formulate some ideas for preventing accidents.

Session A

2 Definition of an accident

'Fatal accident', 'happened accidentally', 'happy accident' – when we use the word 'accident' we wrap up a lot of assumptions about chance, disaster and responsibility, or lack of it. So what **do** we mean by an 'accident'?

Activity 1

What is an accident? Try to define it, in your own words.

_____
_____
_____
_____

There are many ways of explaining the word 'accident', but most people would think of an accident as something that:

- occurs by chance, rather than by design;
- is unwanted, unplanned and unexpected;
- perhaps involves harm or injury to someone;
- often results in interruptions to normal activities, including work;
- perhaps results in damage to physical things, or to the environment.

One definition of an accident is given by the Health and Safety Executive:

Accident includes any undesired circumstances which give rise to:

- ill health or injury;
- damage to property, plant, products or the environment;
- production losses or increased liabilities.[2]

However, it is unwise to ignore incidents which **do not** result in harm, damage or business loss. If you are driving in a line of traffic that suddenly comes to a stop, and you narrowly avoid hitting the car in front, it may make you realize that you aren't leaving yourself sufficient stopping distance. In the same way, the Civil Aviation Authority investigate 'near misses' involving aircraft because they know that doing so will help to prevent serious accidents.

A couple of other words that we will use in this workbook are 'hazard' and 'risk'.

[2] This definition, together with those for 'hazard' and 'risk', is adapted from *Successful Health and Safety Management*, published by HSE Books, Crown copyright, 1991.

Session A

- **Hazard** means **the potential to cause harm**, including:
  - ill health and injury;
  - damage to property, plant, products or the environment;
  - production losses or increased liabilities.

- **Risk** means **the likelihood that a specified undesired event will occur** due to the realization of a hazard:
  - by, or during, work activities; or
  - by the products and services created by work activities.

So, hazard is the potential to cause harm, damage or business loss. Risk is the likelihood that the potential harm from a hazard will be realized.

Activity 2

3 mins

One more definition: how would you explain the word **danger**?

_____
_____
_____
_____

Danger can be described as:

- **an unacceptable level of risk**; or
- **liability or exposure to harm**; or
- **something that causes peril**.

Safety is the result of the activities we carry out to **keep something or somebody from harm**, and could be called the opposite of danger.

We achieve safety by protecting ourselves and our environment, and by identifying, assessing, and then reducing or eliminating, the risk from the hazards we may encounter.

This workbook is mostly concerned with recognizing hazards and preventing accidents. We start with the assumption that:

all accidents are preventable.

Session A

3 What kind of accidents?

> **EXTENSION 1**
> This book is listed on page 94. This list is adapted from page 3 of the book.

In his book *Classic Accidents*, David Farmer list seven 'classic' accident types. These are not the only kind of accident, but are ones that occur repeatedly, in workplaces everywhere. As David Farmer says: 'They keep on happening to different people, at different times, and in different places.' The seven classics are:

- people getting caught in machinery by their hair, or by something they are wearing;
- people falling when using ladders;
- collisions involving reversing vehicles;
- people being burned when using or handling flammable liquids, usually as a result of being ignorant of the hazards;
- people in confined spaces, sometimes together with their rescuers, being overcome by gassing or asphyxiation;
- accidents by tripping, slipping, or falling;
- accidents during maintenance work.

Let's look at some cases.

Entanglements in machinery

> Entanglement accidents often start with a strand of hair, or a loose item of clothing, which becomes caught up. The victim is usually unable to draw back before a finger, a limb, or the whole body, becomes enmeshed.

- A young man was using a portable electric drill fitted with an abrasive disc to clean some metalwork. As he leaned over the work, his tie became entangled and wrapped itself around the drill chuck. Before he could stop the drill, it had wound itself around the tie, and travelled upwards. The abrasive disc caused the young man severe facial injuries.

- A woman using a bench-mounted drill moved her hand across the bench to clear it of clutter. The drill bit caught in her wedding ring, and severely damaged the flesh and tendons of her finger, which surgeons later had to amputate.

Ladder accidents

> Ladders, though useful, are not inherently safe. Even when the ladder is securely fixed, so that it cannot move from side to side, it is still easy for the user to slip on the rungs.

- A man slipped while climbing a short ladder, which was resting on a scaffold tower. Neither ladder nor tower was properly secured, and the tower fell over. The man's head hit a concrete path, injuring him fatally.

- A decorator was climbing a wooden ladder carrying an open pot of paint. His shoes were wet, and one foot slipped on a rung. While he was struggling to regain his balance, the paint splashed over his face and eyes, temporarily blinding him. He again lost his footing and fell, breaking both legs.

Session A

Reversing vehicle accidents

> Visibility from the cab of a reversing lorry is typically poor. Warning devices help, but there is no real substitute for a trained banksman on the ground, telling the driver what to do.

- A delivery lorry, reversing in a school car-park, collided with a car driven by a parent, and rammed it against a wall. Although the lorry was fitted with an automatic reversing horn, the car-driver failed to hear it over the noise of her radio. She was trapped in the vehicle for about an hour, and received severe leg lacerations.

- A refuse collection vehicle was reversing outside a hospital, when it hit an outpatient, killing him outright.

Flammable liquid accidents

> A liquid cannot actually burn, but many types of liquid found in workplaces (solvents, cleaning fluids, paints, fuels) give off vapours that are highly flammable.

- In a hospital surgical unit, a theatre nurse spilled some ether on her uniform. A short while later, a spark was formed from the static electricity generated by her underclothes, which were made from a synthetic material. As a result, the ether-impregnated uniform caught fire, and she was badly burned.

- A man took the bung out of a drum that had been standing around for four years, and applied an oxy-acetylene torch to the drum, with the intention of removing the top. Flammable gases had been accumulating inside the drum, and the torch caused an explosion.

Accidents in a confined space

- A farm labourer was asphyxiated while trying to unblock an underground slurry reception tank. The farmer, in his efforts to rescue his employee, was also overcome. Both men died. Slurry gives off gases that drive out oxygen; the men in the tank were simply unable to breathe.

> Gas can be trapped in a confined space, and may build up slowly and insidiously. When you realize you're in trouble, it is often impossible to escape quickly enough.

- A brewery worker was cleaning the inside of a large copper vessel. Unable to remove some stubborn stains, he decided, without reference to anyone else, to apply nitric acid. Later, he became ill, and was taken to hospital. Doctors were initially given incorrect information about the work he had been doing by the brewery management, who assumed the worker had been using the recommended materials, which would be either a mild alkali solution or a paste made from pumice and tartaric acid. When it was discovered he was suffering from the effects of nitrous fumes, the man was treated appropriately, and fortunately recovered.

Session A

Tripping, falling, or slipping accidents

> Slips, trips, and falls account for around twenty per cent of all accidents at work in the UK. Injuries are often serious, and occasionally fatal.

- A twenty-year old packer went to the tea-room for her midday refreshment, where boiling water for making tea was kept in a free-standing electrical boiler. She slipped on the wet floor, instinctively grabbed the boiler for support and pulled it over on top of her. She received extensive burns. The water on the floor had come from the boiler, which had a dripping tap.

- A woman visitor was walking in the grounds of a sports centre, carrying some equipment. As she walked over a 'sleeping policeman' ramp in the road, the heel of her shoe caught in a circular slot used to carry a fixing bolt. She fell so badly on one arm that a smashed bone had to be removed from her elbow.

Accidents during maintenance

> About a quarter of fatal accidents take place during maintenance operations on buildings or equipment. As this kind of work is usually carried out infrequently, workers are not always familiar with the hazards.

- A young worker was cleaning inside a mixing vessel, at the bottom of which were rotating blades. The interlock system was faulty, and the vessel was powered up while the worker was still inside. The young man lost both his feet, which were cut off at the ankles.

- A self-employed contractor arrived early one morning to start work repairing a school roof. Impatient for his colleagues to arrive, he started to inspect the work to be done, by climbing on, and walking along, the roof. He slipped and fell through a skylight, cutting himself badly.

How can accidents like these be prevented? Before we look at specific accident types, we'll consider the subject in broader terms.

4 What causes accidents?

The 'classic' accident cases we have just looked at were all preventable, but apparently all had different causes. It would be possible, no doubt, to describe fifty or a hundred other work accidents, and each might seem to have come about through a unique set of circumstances.

So how can we talk about 'accident prevention' in any general manner?

Session A

Activity 3

Can you think of **one** feature that **all** work accidents have in common? If you're stuck for an answer look back to the incidents above. Ignore the particular sequences of events that might have led to each accident – and think about **people** and **organizations**.

_____

_____

_____

_____

The 'technical' causes of different accidents may have nothing in common. So to find a common feature we need to move away from technicalities and think in more general terms.

Your response to this Activity may have been to note one of the following:

- poor supervision;
- inadequate training;
- lack of instruction;
- lack of information;
- inadequate procedures.

These would all be correct. The **majority** of accidents result from these kinds of failures.

Although an accident is always unexpected, when we look back we often realize that it should have come as no surprise. Something happens that should never have been allowed to happen. In other words, safety procedures and controls are inadequate.

So perhaps the best response to this Activity might be:

accidents at work are largely caused by poor control and management of safety.

As The Health and Safety Executive (HSE) puts it:

'The majority of accidents and incidents are not caused by 'careless workers', but by failures in control (either within the organization or within a particular job), which are the responsibility of management.'[3]

---

[3] Taken from *Successful Health and Safety Management,* Health and Safety Executive, Crown copyright, 1991, page 9.

7

Session A

Safety at work can only be achieved if there are well-organized **systems** of safety. The reasons are that:

- the safety of people at work depends on co-operation between individuals and between groups;
- people at work largely do what they are told to do or what they are allowed to do;
- people at work tend to assume that someone higher in the organization is making sure that the workplace is safe.

Such a system is a definite policy or strategy for safety, and a set of clear procedures and rules. It takes into account the fact that people are human, are not all equally well-equipped to cope, and may make mistakes. A good system looks at the way people work, and encourages them to be aware of the hazards. When accidents occur, it's usually either because the safety system is inadequate, or because somebody has bypassed the system.

So we can look at reasons why accidents happen from two viewpoints: the circumstances leading up to particular accidents, that is, the technical causes, and the system that allowed them to take place. We can think about accidents on the road from these points of view. The road traffic safety system comprises (at least):

- the state and design of the roads, and programmes of road maintenance;
- the highway code, which sets out the rules of driving;
- the driving test, which is designed to ensure that all drivers reach a minimum standard;
- the drivers themselves;
- the quality of vehicles;
- the policing of roads.

Most road traffic accidents are caused by people doing foolish things (and so bypassing the system), or as a result of mechanical failure. It may be possible to analyse why a particular accident took place, and so do a lot to stop it happening again. But if we want to get the overall accident rate down, we may have to improve the system. (Better roads? Better cars? More stringent policing? Lower speed limits? Different attitudes?)

Later in this session, we will discuss some of the key elements needed for a successful health and safety management system at work.

Let's first look at another couple of accidents and see if we can analyse the causes behind them.

4.1 Who's responsible?

A common first reaction to an accident is to look for someone to blame.

Session A

Activity 4

- Jim Dolling operated a powerful machine with moving parts. Jim was very experienced – he had been using this machine for over three years. The machine came fitted with a safety guard, which prevented anything becoming entangled in the gears. The job involved inserting plastic sheets. Sometimes a sheet got jammed. When this happened, the machine had to be switched off and the guard removed so that the sheet could be extricated. After this had happened a few times, Jim never bothered to replace the guard, because it was a time-consuming operation. His supervisor knew about this, and occasionally used to remind Jim that the guard should be in place. Jim told him not to worry, and in fact Jim never had an accident. Then one day Jim went sick, and someone else took over the machine. This man caught a loose overall strap in the machine and received serious injuries.

1   What were the circumstances that led up to this accident – the technical causes?

   _____

   _____

   _____

2   Who do you think was responsible for the accident? Tick which of the following you think were responsible, and briefly say why this was the case:

   - Jim ☐

   _____

   _____

   - the new operator ☐

   _____

   _____

   - the supervisor ☐

   _____

   _____

   - senior management ☐

   _____

   _____

Session A

- the safety officer – assuming there was one ☐

_____

_____

- the manufacturers of the machine ☐

_____

_____

- the machine maintenance staff ☐

_____

_____

As perhaps you'd agree, the technical causes were that:

- the guard was left off the machine (mainly because it had to be removed frequently and was time-consuming to replace);
- a less experienced operator used an unguarded machine;
- the operator wore loose clothing.

Who was responsible?

- Jim was

  He could – and should – have replaced the guard, especially after being reminded by his supervisor.

- The new operator was

  He should have been more careful about his clothing, and could have replaced the guard before using the machine.

- The supervisor was

  He should have gone further than reminding Jim about the guard – he should have insisted it be replaced. Also, he should not have allowed a less experienced operator to work at the machine without a guard.

- Senior management were

  Responsibility passes all the way to the top of an organization.

- The safety officer, assuming there was such a person in this organization, was **not**

  The essential role of a safety officer is to give advice to the organization and its employees: this job could be compared with that of a quality manager. He or she cannot be expected to monitor every faulty condition that might exist.

- The manufacturers of the machine **perhaps** were

  Did they give enough thought to safety, when they designed a machine with a guard which needed to be removed frequently, and was troublesome to replace?

Session A

- The machine maintenance staff **perhaps** were

    Did they maintain the machine well enough – was it poor maintenance which caused the frequent jams?

> 'If only' is a phrase often heard after an accident.

It may seem that there are a lot of people involved, and this isn't surprising, because activities at work nearly always involve a number of people. The fact is that the accident might well have been avoided **if only** one of these people or groups had done a better job.

4.2 What went wrong?

If something serious happens, it is both natural and useful to investigate what went wrong.

Here's a description of another incident.

Activity 5

5 mins

- Billy had just started work for his local council on a job training scheme and was to spend his time working for the Parks and Highways Department. On his first day, Billy was given overalls and safety boots by the supervisor, and told to wear them whenever he was at work. Billy disliked the idea of wearing boots, and anyway the boots pinched his toes. The next day he came to work in his favourite training shoes. In the afternoon, Billy worked with the supervisor lifting some barrels from an open wagon. One of the barrels was so heavy they dropped it. It fell on Billy's foot, breaking all his toes.

This time, imagine you were given the job of investigating this accident. Jot down **two** or **three** important questions you might ask during your investigation.

_____
_____
_____
_____
_____

Session A

> We'll be looking at accident investigation in Session D.

Some of the questions you might have asked are listed here.

- Was Billy told *why* it was important to wear the safety boots, and was it emphasized that he wouldn't be allowed to work unless he was wearing them?
- Were other people, doing similar jobs, allowed to work in their own footwear? In other words, were the rules normally insisted upon?
- Did the supervisor notice that Billy wasn't wearing the boots?
- Was it made clear to Billy that if the boots or overalls didn't fit properly he could ask for a different size?
- How was it that the barrel was heavier than expected? Should a fork-lift truck or other equipment have been used to unload the wagon?
- Had Billy (or the supervisor) been given training in lifting heavy weights?
- Were written instructions provided?

You may well have thought of other questions. In an accident investigation like this, the answers frequently lead to the conclusion that more than one factor is involved.

So we can say that:

- there is often more than one person or group at fault in any accident;
- accidents nearly always have more than one cause.

Another interesting point is that, in both the incidents described in the last two Activities – as with most accidents – the hazards seemed to be perfectly obvious. Anyone operating a machine without a guard is clearly in danger of machine entanglement, and anyone not wearing the proper protective clothing risks being hurt. However, it should be said that not everyone perceives risks in the same way. For one thing, people will often make the mistake of assuming that 'it won't happen to me' – until it does.

Of course, however good the system, it isn't possible to eliminate every hazard. But where hazards are known, it is important to protect people against them.

5 Policies for safety

So what is the best management approach to accident prevention?

Those organizations that have succeeded in improving their safety records have found that a successful approach to safety entails:

- issuing a clear **health and safety policy** that everyone knows and understands;
- establishing safety **objectives** that are realistic and achievable;
- providing **resources** to make the objectives achievable;
- setting safety **standards** that are measurable, and against which the objectives can be compared;

Session A

- **identifying hazards** and **assessing the risks** from them;
- putting systems in place that **eliminate the hazards** or **reduce the risks**;
- setting up procedures for **monitoring** health and safety performance;
- providing systems designed to increase safety **knowledge, awareness, and understanding**, and to encourage people to accept **responsibilities**.

5.1 The health and safety policy statement

This document is the starting point for all accident prevention and health promotion.

Under Section 2(3) of Health and Safety at Work etc. Act 1974, every employer has a duty to:

'prepare, and as often as appropriate, revise, a written statement of his general policy, with respect to the health and safety at work of his employees and the organization, and arrangements for the time being in force for carrying out that policy, and to bring the statement and any revision of it to the notice of his employees'.

Activity 6

Your organization's health and safety policy statement should be available for you to read. Obtain a copy of it, and make sure you understand it. When you have done so, summarize briefly below your own responsibilities under this policy.

_____
_____
_____
_____
_____
_____
_____
_____
_____

Session A

The Health and Safety Executive (HSE) says:

'The best health and safety policies are concerned not only with preventing injury and ill health (as required by health and safety legislation), but also with positive health promotion which gives practical expression to the belief that people are a key resource . . .

The ultimate goal is an organization in which accidents and ill health are eliminated, and in which work forms part of a satisfying life, contributing to physical and mental well-being, to the benefit of both the individual and the organization.'[4]

A good health and safety policy statement will reflect these high aims. It should state the organization's health and safety objectives, and commit all managers in the organization to these objectives. It should make clear:

- management's intentions;
- how the organization is structured to implement the health and safety policy;
- the safety rules, procedures and other arrangements;
- the individual responsibilities of each level of management;
- the role of health and safety specialists, such as safety officers, advisers, company doctors and so on.

So your health and safety responsibilities, and those of your colleagues, should be clearly identifiable in your organization's policy statement.

6 The team leader's role

How does the local team leader fit into this picture?

Activity 7

6 mins

Now that you have read and understood your organization's health and safety policy statement, how do you think you, as a team leader, can contribute in practical ways to your organization's safety policy?

_____
_____
_____
_____
_____
_____

[4] Taken from *Successful Health and Safety Management*, Health and Safety Executive, Crown copyright, 1991, pages 5 and 6.

Session A

There are no doubt a number of ways of contributing, including perhaps:

- **working with your manager** to agree and maintain standards of safety in your work area;
- **communicating with your team** about safety, explaining the rules, answering questions and establishing that your team members know what is expected of them;
- **developing a good team spirit**, so that efforts to maintain and improve safety standards are made in unison;
- **using all the resources** you have in a safe as well as an efficient manner;
- **getting individuals** in your team **to accept responsibility** for their own safety and for that of others;
- encouraging your team to **report all incidents**, including near misses.

In summary, this means getting the team to **work together to prevent accidents** and to **share responsibility for safety**.

We will return to this theme, later in the workbook.

Self-assessment 1

1 Match each term on the left with the correct definition, chosen from the list on the right.

Hazard	Any undesired circumstances which give rise to ill health or injury; damage to property, plant, products or the environment; production losses or increased liabilities.
Danger	The potential to cause harm, including ill health and injury; damage to property, plant, products or the environment; production losses or increased liabilities.
Accident	The likelihood that a specified undesired event will occur due to the realization of a hazard by, or during, work activities; or by the products and services created by work activities.
Safety	An unacceptable level of risk.
Risk	The result of the activities we carry out to keep something or somebody from harm.

2 Complete the following three statements, by selecting the most correct ending from the list on the right.

a The seven 'classic' accident types
   i cover all the known accidents at work.
   ii could all be blamed on inadequate training.
   iii keep on happening to different people, in different places.
   iv are mostly the result of foolish people doing foolish things.

Session A

b Accidents at work are largely caused by
    i ignorance.
    ii safety systems out of control.
    iii technical factors.
    iv people doing what they're told to do, without thinking about the consequences.

c An organization's health and safety policy statement
    i is the starting point for all accident prevention and health promotion.
    ii consists of a list of safety rules.
    iii is confined to higher management.
    iv is a technical document required by the law, but which few people read.

Answers to these questions can be found on page 96.

7 Summary

- An accident is any undesired circumstances which give rise to ill health or injury; damage to property, plant, products or the environment; production losses or increased liabilities.
- A hazard is the potential to cause harm, including ill health and injury; damage to property, plant, products or the environment; production losses or increased liabilities.
- Risk is the likelihood that a specified undesired event will occur due to the realization of a hazard by, or during, work activities; or by the products and services created by work activities.
- Danger is an unacceptable level of risk.
- Safety is the result of the activities we carry out to keep something or somebody from harm.
- All accidents are preventable.
- Accidents at work are largely caused by safety systems out of control.
- Safety at work can only be achieved if there are well-organized systems of safety.
- There is often more than one person or group at fault in any accident.
- Accidents nearly always have more than one cause.
- The health and safety policy statement is the starting point for all accident prevention and health promotion.
- The team leader can contribute to the organization's health and safety policy by getting the team to work together to prevent accidents and to share responsibility for safety.

Session B The management of safety

1 Introduction

It is worth repeating the HSE statement we read earlier:

'**The majority of accidents** and incidents are not caused by 'careless workers', but by **failures in control** (either within the organization or within a particular job), which are **the responsibility of management**.'[5]

Managers are responsible for creating health and safety policies, and for managing and supervising people to implement safe systems of work.

This session is devoted to a number of aspects of management related to safety, including: strategies, the law, risk assessment, and health and safety committees.

We begin with a subject very close to any manager's heart: costs.

2 The cost of accidents

EXTENSION 2
The HSE booklet *The Costs of Accidents at Work* describes five case studies, analysing the cost of accidents for each. It is worth getting hold of a copy if you are concerned about costs.

The cost of accidents and health problems at work can be measured in financial terms, both to the employer and to the injured or sick person.

Let's look at the employer's position first.

---

[5] Taken from *Successful Health and Safety Management*, Health and Safety Executive, Crown copyright, 1991, page 9.

Session B

Activity 8

Spend a few minutes writing a list of the ways in which an accident at work may cause an employer to lose money. Try to think of **three**.

Money may be lost by the employer through:

- not having the services of the injured person while he or she is unable to work, including perhaps the cost of hiring temporary staff;
- the disruption to the work of other people;
- the time spent by the supervisor in training replacement staff and perhaps taking part in an accident enquiry;
- possible damage to equipment;
- cost of employee welfare benefits;
- possible claims for compensation.

Financial losses can be separated into insured and uninsured costs.

EXTENSION 2
This diagram has been redrawn from page 9 of this booklet.

The following diagram shows one example of the hidden costs of accidents – the accident iceberg – found during an HSE study of a creamery owned by a large multinational company.

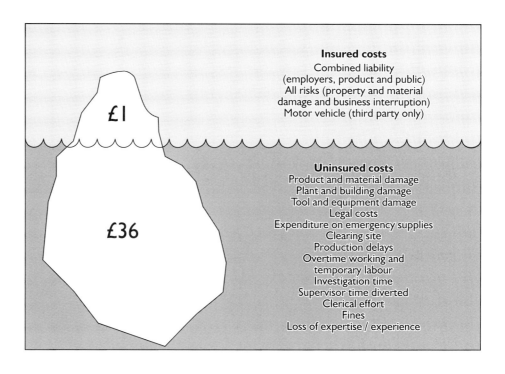

Session B

Of course, the extent of insured losses, for any particular organization, depends upon the kinds of insurance policy held. The ratio of 1:36 does not seem to be untypical, however.

How about the employee's financial position?

Activity 9

3 mins

Note how an employee may lose money as a consequence of having had an accident at work.

_____

_____

_____

Here are a few suggestions.

- Even with sickness benefit, the person's income may go down, especially if the injury results in a long period of absence from work.
- In the worst case, he or she may not be able to return to work at all.
- There may be other costs – travel to and from a medical centre, costs of prescriptions and so on.

Cumulatively, accidents and health problems have an effect on the national economy. The worse the health and safety record of work organizations, the worse off we all are.

The costs in human terms – the physical and mental distress of the people involved and their families – are much more difficult to evaluate, although just as real. If you've known someone who has been badly hurt at work, you'll appreciate how high the human costs can be.

3 Management strategies for safety

As Jeremy Stranks suggests in his book *Management Systems for Safety*, an organization can adopt one of a number of approaches to safety. It can:

EXTENSION 3
This book is listed on page 94. These points have been paraphrased from pages 3 and 4 of the book.

- do just enough to keep within the law;
- consider safety in purely financial terms, balancing the costs of accidents against the costs of implementing safety policies;
- decide that its employees are valuable assets, and therefore worth protecting;
- take what is sometimes called a 'human factors' approach to safety.

Let's look at what we mean by the last of these.

19

Session B

3.1 The human factors approach

To develop the human factors approach, the organization must put in place a strategy based upon:

- a positive safety culture, in which health and safety is recognized as being crucial to the realization of the organization's aims;
- systems that take account of employees' varying capabilities, while acknowledging that nobody is infallible;
- a commitment to high standards through all levels of management, starting at the top.

Safety strategies need to be **proactive**, rather than reactive: that is, they should control the situation by taking the initiative, rather than simply reacting to what happens. Both workplaces and people need to be made safe.

3.2 Making workplaces safe

To carry out work, organizations typically use premises, equipment, processes and materials. So:

- **premises** must be safe

  Buildings must be stable and sound, and must offer employees protection;

- **equipment** must be safe

  It should be appropriate to its task, and properly maintained;

- **processes** must be safe

  Every aspect of a work process has to be considered in terms of its hazards and risks;

- **materials** must be safe

  The hazards from substances need to be recognized, and the risks contained.

Most importantly, **safe systems of work** must be employed integrating all of the above. A safe system of work can be defined as:

the integration of people, machinery and materials in a correct working environment to provide the safest possible working conditions.

Session B

3.3 Making people safe

Things can still go wrong, even in an apparently safe workplace because people are unpredictable, vulnerable and may be ignorant of the dangers.

Activity 10

So far, we have listed safe premises; equipment, process, and materials. What else is needed in a workplace to help keep people safe? [Hint: think about your own job as first line manager.]

_____

_____

> In all matters of safety, it's important to take into account the capabilities and needs of individuals. It can be a serious mistake to assume that everyone will behave in the same way in response to hazards and risks.

An important ingredient for workplace safety is **adequate supervision**. People need different amounts of supervision, according to their age, experience, and status. Certain groups are more vulnerable, particularly:

- the young;
- people with disabilities;
- workers exposed to special risks.

With regard to the last of these, **personal protective equipment (PPE)** often has a significant role to play. We will look at the various types of PPE in the next session.

In many workplaces, adequate **washing facilities** must be provided, especially where:

- there is a risk of hazardous chemicals coming into contact with the skin;
- food is handled.

3.4 Training

A key aspect of accident prevention is training. At an absolute minimum, people need training in:

- understanding health and safety law;
- using personal protective equipment;
- specific hazards in their workplace, and how to deal with them;
- dealing with emergencies and abnormal occurrences;
- safe and correct operation of machinery;
- manual handling techniques (if applicable).

Training should (by law):

- be repeated periodically where appropriate;
- be adapted to take account of new or changed risks;
- take place during working hours.

Session B

 ## Activity 11

This Activity may provide the basis of appropriate evidence for your S/NVQ portfolio. If you are intending to take this course of action, it might be better to write your answers on separate sheets of paper.

What training have you arranged for your team in health and safety, in the past year?

Explain how you ensure that the training your team receive is appropriate and sufficient for them to work in a healthy, safe and productive way.

What health and safety training are you planning for the coming months?

Session B

4 Accident prevention and the law

At the very least, organizations must comply with the law on health and safety. The penalties and costs of not doing so can be very high, including fines and imprisonment.

In brief, health and safety legislation is embodied in:

> This section gives just a brief summary of the law related to accident prevention. The Super Series workbook *Managing Lawfully – Health, Safety and Environment*) contains more information on this subject.

- case law or common law

    Every court is bound by the decisions previously made in higher courts, or courts at the same level. Thus a judgement made in a particular case can affect the outcome of all subsequent cases of the same kind.

- statute law

    Statute law comprises **Acts of Parliament**, such as the Health and Safety at Work etc. Act, 1974 (HSWA). 'Enabling Acts' like HSWA give rise to 'statutory instruments', in the form of **Regulations**, such as the Management of Health and Safety at Work Regulations, 1992 (MHSWR).

- European Union directives

    Directives are the instruments of EU legislation, which are binding on all member states. Normally, directives are complied within the UK by embodying them in Acts of Parliament.

We will take a brief look at both HSWA and MHSWR.

4.1 The Health and Safety at Work etc. Act, 1974 (HSWA)

The most significant and far-reaching piece of legislation, covering accidents and health in workplaces generally, is the Health And Safety at Work etc. Act 1974 (HSWA).

> EXTENSION 4
> Important as they are, it would be wrong to assume that HSWA and MHSWR are the sum total of health and safety legislation. In this extension you can see a list of 11 Acts and 32 sets of Regulations that are concerned with health and safety law.

The purpose of HSWA is to:

'provide the legislative framework to promote, stimulate and encourage high standards of health and safety at work'.

HSWA places obligations on both employers and employees, and covers the safety of everyone at work.

Employers have to safeguard the health, safety and welfare at work of all their employees,

'so far as is reasonably practicable'.

Session B

To illustrate the use of the term 'reasonably practicable', imagine a chemical engineering works, where the products are hazardous: flammable and toxic chemicals. The chemical company would be expected to safeguard the health of its employees, and one important step would be to issue operators with protective clothing. It wouldn't be 'reasonably practicable', however, to say that chemical workers should be protected to the extent of never having to deal with hazardous chemicals.

Employers can and should provide:

- safe ways of working;
- safe equipment and machinery;
- safe premises;
- safe handling, storing and transporting of goods;
- information;
- instruction;
- proper training and supervision.

As a first line manager, you are often required to act on behalf of your employer. The last three items on the list above – information, instruction, and proper training and supervision – are therefore particularly significant for you.

Under HSWA, employers must also have regard for the safety of non-employees – members of the public, contractors' staff and so on – who may be affected by the activities of their companies.

Failure to implement this part of the Act could lead to managers and supervisors being prosecuted.

Employees also have duties under HSWA.

Activity 12

What duties do you think employees have for health and safety at work?

_____

_____

_____

_____

All employees have duties:

- to take reasonable care for their own safety and that of others;
- to co-operate with their employers in matters of safety;
- not to interfere with or misuse anything provided for safety.

Session B

4.2 The Management of Health and Safety at Work Regulations, 1992 (MHSWR)

In January 1993, six new important Regulations (the 'six-pack' Regulations) came into force, which are all relevant to most workplaces. They are the:

- Management of Health and Safety at Work Regulations, 1992 (MHSWR);
- Workplace (Health, Safety and Welfare) Regulations, 1992 (WHSWR);
- Personal Protective Equipment at Work Regulations, 1992 (PPEWR);
- Health and Safety (Display Screen Equipment) Regulations, 1992 (HSDSER);
- Provision and Use of Work Equipment Regulations, 1992 (PUWER);
- Manual Handling Operations Regulations, 1992 (MHOR).

We will only examine the contents of MHSWR in this workbook, as it is more far-reaching than the others.

> **EXTENSION 3**
> This quote is from page 14 of this book.

MHSWR clarifies many of the points made in HSWA. However, it goes further than this. In *Management Systems for Safety*, Jeremy Stranks notes:

'Managers must appreciate that the room for manoeuvre is drastically reduced under these Regulations in the light of the absolute nature of their requirements. It will not be possible to plead, when charged with an offence, that it was not "reasonably practicable" in the circumstances under consideration.'

> **MHSWR applies to every workplace except sea-going ships.**

Whereas HSWA gives organizations a great deal of scope in implementing health and safety policies, MHSWR spells out more precisely what must be done.

MHSWR requires employers in virtually all work activities in Great Britain and offshore to:

- assess the risks of the job

> **We will look at risk assessment shortly.**

The employer must make an assessment of the health and safety risks of work activities to employees and to anyone else who may be affected, and record the findings of this assessment. Following this, arrangements have to be made to put any necessary risk-reducing measures into practice.

- provide health surveillance if necessary

Where risk assessment shows it to be necessary, employers have to provide appropriate health surveillance for employees.

Session B

Activity 13

In what kind of industries would employees need to have their health surveyed?

_____
_____
_____
_____

Health surveillance is particularly relevant to employees who work in hazardous conditions, and those who work with poisonous substances, such as chemicals and dusts.

Perhaps you included your own industry in your list. Surveillance of health may include examining medical records and making medical examinations. It might also require supervisors and managers to take careful note of, and to act on, any comment or complaints about health by employees. Every employee may need to be warned to watch out for particular health conditions, such as skin irritation.

MHSWR also requires employers to:

> We will look at emergency procedures in Session D.

- appoint competent people (if staff are not trained and competent to deal with the requirements of MHSWR, external consultants may need to be appointed);
- provide employees with information and training about health and safety matters;
- set up emergency procedures.

For their part, employees are required to follow health and safety instructions and to report hazards.

5 Risk assessment

MHSWR, and four of the other 'six-pack' Regulations listed on page 25, require organizations to assess the risks attached to work operations.

According to MHSWR, an employer must assess:

- 'the risks to the health and safety of his employees to which they are exposed at work';
- 'the risks to the health and safety of persons not in his employment arising out of or in connection with the conduct of him and his undertaking'.

Session B

A definition of risk assessment is:

> **EXTENSION 3**
> This definition is from page 54 of this book.

'an identification of the hazards present in an undertaking and an estimate of the extent of the risks involved, taking into account whatever precautions are already being taken'.

The process is one of:

- identifying the hazard;
- measuring and evaluating the risk from this hazard;
- putting measures into place that will either eliminate the hazard, or control it.

The risk assessment has to be systematic, so that no risks are overlooked. Organizations may approach the task in various ways. They could, for example, examine:

- every activity that might result in injury;
- every type of substance used;
- every type of machine used;
- every location on which work takes place.

Each of these approaches is valid, provided that it is carried out systematically, and covers all possible risks.

5.1 Calculating the risk

The assessment of risk is not an exact science, because it involves making judgements about:

- the likelihood or probability that an accident might occur;
- how serious the outcome might be, if it did occur;
- how often the risk is present.

We will briefly discuss two methods for calculating risk.

> **You are not advised to apply these methods to your work situation without further training.**

One method is to simply rate, first the hazard severity and second, the likelihood of occurrence.

Session B

This is the first method of calculating risk that we discuss.

Hazards – the potential to cause harm will vary in severity. The effect of a hazard may, for example be rated as:

3 – major: resulting, for example, in death or major injury;

2 – serious: causing people to be off work for more than three days;

1 – slight: all other minor injuries.

Harm may not arise from exposure to a hazard in every case and in practice the likelihood of harm will be affected by the organization of the work, how effectively the hazard is controlled, and the extent and nature of exposure to it.

The likelihood of harm may be rated as:

3 – high: where it is certain or near certain that harm will occur;

2 – medium: where harm will occur frequently;

1 – low: where harm will seldom occur.

The risk can then be defined as the combination of the severity of the hazard with the likelihood of its occurrence, so that:

$$\text{RISK} = \text{HAZARD SEVERITY} \times \text{LIKELIHOOD OF OCCURRENCE}$$

The single figure result (ranging from 1 to 9) provides a method of comparing the risk associated with various work operations.

> **Example.**
>
> The risk to health from the spillage of a hazardous chemical in a company works is calculated as follows:
>
> hazard severity – major (rating 3)
>
> likelihood of occurrence – low (rating 1)
>
> Risk = Hazard Severity $\times$ Likelihood of Occurrence = $3 \times 1 = 3$

This simple method may not suit all organizations, and other methods have been devised.

Session B

This is the second method of calculating risk that we discuss.

A second method combines the three factors of:

- likelihood (probability) of occurrence;
- severity; and
- frequency.

A scale from 1 to 10 is assigned for each one, and the three scores are multiplied together to get a rating out of 1000.

The formula is:

> Risk rating = Probability (P) × Severity (S) × Frequency (F)

The following tables are used.

Probability scale

Probability index	Descriptive phrase
10	Inevitable
9	Almost certain
8	Very likely
7	Probable
6	More than an even chance
5	Even chance
4	Less than an even chance
3	Improbable
2	Very improbable
1	Almost impossible

Session B

Severity scale

Severity index	Descriptive phrase
10	Death
9	Permanent total incapacity
8	Permanent severe incapacity
7	Permanent slight incapacity
6	Absent from work for more than 3 weeks with subsequent recurring incapacity
5	Absent from work for more than 3 weeks but with subsequent complete recovery
4	Absent from work for more than 3 days with subsequent complete recovery
3	Absent from work for less than 3 days with complete recovery
2	Minor injury with no lost time and complete recovery
1	No human injury expected

Frequency scale

Frequency index	Descriptive phrase
10	Hazard permanently present
9	Hazard arises every 30 seconds
8	Hazard arises every minute
7	Hazard arises every 30 minutes
6	Hazard arises every hour
5	Hazard arises every shift
4	Hazard arises once a week
3	Hazard arises once a month
2	Hazard arises every year
1	Hazard arises every 5 years

Session B

> **Example.**
>
> Suppose a certain work operation involved a hazard which:
>
> - would probably cause an accident;
> - would result in an absence from work for more than three weeks, with subsequent recovery;
> - arose once an hour.
>
> By looking up the tables, we would find that P = 7, S = 5 and F = 6. Applying our formula, we would get:
>
> Risk rating = Probability (P) × Severity (S) × Frequency (F).
>
> = 7 × 5 × 6 = 210
>
> Next, we apply the risk rating to a table showing the priority of action:
>
> **Priority of action scale**
>
> | 801–1000 | Immediate action |
> |---|---|
> | 601–800 | Action within 7 days |
> | 401–600 | Action within next month |
> | 201–400 | Action within next year |
> | Below 200 | No immediate action necessary, but keep under review. |
>
> For our example, the priority would be 'action within the next year'.

Activity 14

3 mins

Assume that you found the following situation during a risk assessment. You determined that there was more than an even chance of an accident occurring, that there would be permanent slight incapacity if it did occur, and that the hazard arises every 30 minutes. What is the risk rating, and how quickly would action need to be taken?

Session B

Compare your answer with the following.

> Risk rating = Probability (P) × Severity (S) × Frequency (F).
>
> = 6 × 7 × 7 = 294

According to the priority of action scale, action would need to be taken within a year.

The methods described above have their drawbacks, and need to be administered with care. If your organization uses procedures for risk calculation, you may be able to get training in their application.

5.2 Taking action

Risk assessment is an important management activity, whether or not it is quantified on a scale. When hazards are present, action must be taken to assess the risk and eliminate or reduce it.

Activity 15

This Activity may provide the basis of appropriate evidence for your S/NVQ portfolio. If you are intending to take this course of action, it might be better to write your answers on separate sheets of paper.

As part of your job, you are expected to ensure that the work conditions under your control conform to organizational and legal requirements. In fulfilling this duty, you may have to take part in risk assessments. You should, at the least, be aware of hazards present in your work area, so that you can pass information about them to your team.

Give **three** examples of identified hazards that you are aware of in your work.

_____
_____
_____
_____
_____
_____

Session B

What are the risks associated with these hazards, according to the last risk assessment?

_____
_____
_____
_____
_____
_____
_____

Describe any further actions you plan to take, and when, in order to eliminate or reduce the risk from these hazards.

_____
_____
_____
_____
_____
_____
_____

Ideally, risks should be eliminated completely. You might ban the use of a particular harmful substance, for example. Or you could try to find a different, less hazardous, method of doing the work.

If **elimination of the risk** is not practicable, other measures you might take, in order of preference, are to:

1 **enclose the risk**, say by placing a guard around a machine, or putting a hazardous chemical in a suitable container, to prevent anyone coming in contact with it;

2 **install a safety device**, such as an interlock that precludes access to a device unless the power is off;

3 implement a safe system of work, so that the **risk is reduced to an acceptable level**.

If there is still a risk present, you may need to take other measures.

4 You could **provide specific written safety instructions** for your team, and make sure they understand what the risks are, and that they know how to protect themselves against the risks.

Session B

> If you have team members whose first language is not English, it may be wise to provide translations of safety instructions, whether or not they request this.

5 As we have already discussed, **safety supervision** must be adequate for the people you are responsible for. Also, even if you believe that your written instructions are clear, a face-to-face discussion can help to clear up any misunderstandings. Not everyone is a careful reader.

6 Everyone at work needs health and safety **training**. Training may be in respect of particular hazards, or may cover a number of areas.

7 **Information** should be provided, where hazards are known to exist, in the form of posters, safety signs, warning notices and so on.

8 As a last line of defence, your team may need to have **personal protective equipment** such as goggles, helmets, aprons and so forth.

6 People with a special role to play

Everyone at work has a part to play in health and safety.

But these subjects are too important to be left to compete for priority with all the other activities going on in a busy organization.

It's also valuable to have people who, as part of their job, deliberately and consciously set out to promote health and safety and to look into potential hazards.

6.1 The safety officer

Full- or part-time **safety officers** (also sometimes called safety practitioners, safety specialists, safety advisers or safety engineers) may be appointed.

A safety officer is part of the management team, directing employees in safety activities and sharing the responsibility with the other managers and supervisors. This may not necessarily be a full-time task. The job may be delegated to a number of people, each having an assistant's role for a specialized activity.

> **EXTENSION 5**
> If you are interested in studying the subject of safety representatives or safety committees in more depth, the Approved Code of Practice, *Safety Representatives and Safety Committees*, by the Health and Safety Executive, gives useful information.

Because the officer is specially trained and/or qualified in safety matters, he or she will be in a position to advise others.

The safety officer co-ordinates and promotes the work of accident prevention by:

- periodically inspecting plant, tools and machinery to identify hazards;
- examining work operations to determine whether there are any unsafe practices;
- recommending improvements;
- taking part in safety training and education of employees and supervisors;
- acting as co-ordinator of safety work;
- leading and taking part in safety meetings.

Session B

6.2 The safety representative

Recognized trade unions may appoint safety representatives to represent employees on matters of health and safety at work.

A safety representative is not necessarily a safety specialist like the safety officer. Nevertheless, the representative is expected to have worked for his or her present employer for at least the preceding two years, or have had two years' experience in a similar job.

Activity 16

3 mins

Although the representative is not expected to be a safety specialist, he or she is in a good position to contribute certain kinds of knowledge when it comes to accident prevention.

What knowledge might an experienced employee, interested in safety, be able to contribute?

_____
_____
_____

Such a person may have knowledge of:

- the hazards in that type of work;
- the specific hazards in that workplace;
- what has been tried in the past, successfully or unsuccessfully, to eliminate hazards in that workplace;
- accidents that have happened in the past;
- the health and safety policy of the employer, and the arrangements for carrying out the safety policy.

Safety representatives are of course expected to keep such knowledge up to date. They are also expected to keep themselves informed of the legal requirements related to health and safety – particularly those of the groups of people they represent.

The employer must give safety representatives time off with pay for training in safety matters.

Session B

Activity 17

We have established that a safety representative is experienced, interested in safety, receives training and represents employees in health and safety matters.

With this in mind, what specific functions should the representative carry out, do you think? You should try for two or three.

_____
_____
_____
_____

> The relevant legislation is The Safety Representatives and Safety Committees Regulations 1977.

The following are the main functions of the safety representative in law. They are probably what you would expect. The law (Safety Representatives and Safety Committees Regulations, 1977) says that the representative shall:

- 'investigate potential hazards and dangerous occurrences at the workplace – whether or not they are drawn to his attention by the employees he represents – and examine the causes of accidents at the workplace';
- 'investigate complaints by any employee he represents, related to that employee's health, safety or welfare at work';
- 'make representations to the employer on specific or general matters of health, safety and welfare';
- 'carry out inspections';
- 'attend meetings of safety committees'.

The employer must let the representative have time to carry out this work, without loss of pay.

Of course, in practice, people like yourself, who have direct responsibility for the safety of staff, will normally work closely with safety representatives when it comes to inspections and investigations.

Full co-operation is essential in health and safety matters.

6.3 Safety committees

In law, an employer must set up a safety committee if requested to do so by two or more safety representatives. In practice, safety committees are established quite frequently by employers even without such a request.

To be most effective, safety committees should be established for each place of work, rather than having one committee trying to serve a large organization with lots of sites.

Session B

The work of a safety committee might include:

- studying statistics and trends;
- considering reports by safety representatives;
- helping to develop safety rules and safe systems of work;
- monitoring the effectiveness of safety training;
- monitoring the effectiveness of communication and publicity on health and safety;
- making recommendations to management to improve health and safety at work in both a specific and general way.

Activity 18

Which group or groups do you think should be represented on a safety committee?

Employees? ☐

Supervisors/ managers? ☐

Senior management? ☐

A safety committee should have people representing employees and managers at all levels.

If there is a safety officer, he or she should take part. Any medical staff appointed by the organization would also have a role to play. Technical specialists may be brought in when specific subjects are being discussed.

Self-assessment 2

1 Which two of the following statements are correct?

   a Uninsured costs are typically much lower than insured costs. ☐

   b Proactive safety strategies, rather than reactive ones, are best. ☐

   c It is worth remembering that nearly everyone will react to a particular hazard in the same way. ☐

   d Safe processes, safe premises, and safe materials, are all necessary and important in making workplaces safe. ☐

Session B

2 Complete the following sentences with suitable words, chosen from the list below.

   a A safe _____ of work is the _____ of people, machinery and materials in a correct working _____ to provide the safest possible working _____ .

   b Under HSWA, _____ must have regard for the safety of _____ who may be affected by the activities of their companies.

   c All _____ have duties: to take _____ care for their own safety and that of others; to _____ with their employers in matters of safety; not to _____ with or misuse anything provided for safety.

   d Every _____ must make an _____ of the health and safety _____ of work activities to _____ and anyone else who may be affected, and record the findings.

   e MHSWR requires employers to appoint _____ people.

ASSESSMENT	CO-OPERATE	COMPETENT
CONDITIONS	EMPLOYEES	EMPLOYEES
EMPLOYER	EMPLOYERS	ENVIRONMENT
INTEGRATION	INTERFERE	NON-EMPLOYEES
REASONABLE	RISKS	SYSTEM

3 Explain, briefly, the purpose of assigning risk ratings to hazards.

   _____
   _____
   _____
   _____

4 Place the following actions in order of preference, as responses to a known risk from a hazard, by putting a number in each box. (The most preferable action should be assigned number 1.)

   Implement a system of work that reduces the risk. ☐
   Install a safety device. ☐
   Eliminate the risk. ☐
   Provide specific written safety instructions. ☐
   Enclose the risk. ☐
   Provide training. ☐
   Provide general information about safety. ☐
   Provide personal protective equipment. ☐
   Supervise those at risk from the hazard. ☐

   The answers to these questions can be found on page 97.

Session B

7 Summary

- In terms of financial cost, employers and employees both suffer when accidents occur. Frequently, employers' costs are far greater than those against which they are insured.

- Safety strategies need to be proactive, rather than reactive: that is, they should control the situation by taking the initiative, rather than simply reacting to what happens.

- For a workplace to be safe, premises, equipment, processes and materials must all be made safe.

- A key aspect of accident prevention is training.

- The Health and Safety at Work etc. Act, 1974 (HSWA) places obligations on both employers and employees, and covers the safety of virtually everyone at work. Employers must safeguard the health, safety, and welfare of their employees, and of non-employees affected by work activities.

- HSWA says that all employees have duties:
  - to take reasonable care for their own safety and that of others;
  - to co-operate with their employers in matters of safety;
  - not to interfere with or misuse anything provided for safety.

- The Management of Health and Safety at Work Regulations, 1992 (MHSWR) requires employers to:
  - assess the risks of the job;
  - provide health surveillance if necessary.
  - appoint competent people;
  - provide employees with information and training about health and safety matters;
  - set up emergency procedures.

- Risk assessment is 'an identification of the hazards present in an undertaking and an estimate of the extent of the risks involved, taking into account whatever precautions are already being taken'. Two methods for obtaining a numerical value for risk rating were discussed.

- To control risks and hazards, the following actions may be taken, in order of preference:
  1. eliminate the risk;
  2. enclose the risk;
  3. install a safety device;
  4. implement a system of work that reduces the risk;
  5. provide specific written safety instructions;
  6. supervise those at risk from the hazard;
  7. provide training;
  8. provide general information about safety;
  9. provide personal protective equipment.

- People with a special role to play in health and safety include safety officers, safety representatives and safety committees.

Session C Practical accident prevention

1 Introduction

> **EXTENSION 6**
> The HSE book *Essentials of Health and Safety at Work* identifies many types of hazard, and the precautions that may be used to protect against them.

We began by looking at some accounts of actual accidents, and tried to establish ways of preventing them. Mostly so far, we have talked in terms of management systems. This is right and proper, for accidents are mainly the result of systems out of control.

But there is plenty to be learned by examining the reasons behind particular accidents and accident types. That's what accident investigators do: they want to know 'how did this accident happen, and how can we stop it happening again?'

So, in this session, we'll put accidents into groups, and see what can be done by way of preventative action. The groups are:

- machinery accidents;
- electrical hazards;
- falling accidents, including accidents with ladders;
- manual handling accidents;
- maintenance accidents;
- fire hazards.

We also look at personal protective equipment and, finally, at housekeeping.

2 Machinery safety

These days, it's hard to think of a job that doesn't involve equipment of some kind.

We normally associate machinery safety with heavy industry or agriculture. But machines are found in offices, shops, studios, playschools, hospitals – in fact in almost every work area.

You may not have had to deal with an accident involving machinery – yet. That doesn't mean to say that all the machines under your responsibility are completely safe, or that everyone who uses them is capable of using them safely under all conditions.

Session C

Activity 19

Think about the machines in your own place of work. What do you do to try to make sure that people don't have accidents when they are in use?

_____
_____
_____
_____

What more could you do?

_____
_____
_____
_____

Here are some general guidelines that apply in all situations.

- The first important thing is to do what you just did – think about how to prevent injury. Think what could occur if things were to go wrong. Learn to recognize when parts of a machine are dangerous.

  It's useful to think about what might go wrong.

  As we discussed in Session A, a very common type of machinery accident is entanglement. Any machine with rotating parts is potentially dangerous, whether or not those parts are projecting or normally accessible.

- Don't rely on training, and on everyone being sensible. Many accidents arise because people behave in a foolish or careless manner.

  Accident prevention shouldn't depend on people always obeying the rules.

- Don't allow guards to be removed. Guards are there to protect people. If guards need to be removed often, they should be interlocked, so that the machine can never be turned on accidentally when the guard is absent. If the guards are inconvenient to use, or are easily defeated, try to get them improved if you can.

  Keeping the guards in place is not just good practice – it's the law.

- Ensure that regular inspection and maintenance is carried out by competent and trained staff.

  Inspection and maintenance of equipment are essential.

Session C

- But remember that maintenance staff get hurt, too. It's important to give them all the information you can before they start work.

- Allow only trained and authorized staff to use the equipment.

Keep untrained and unauthorized people away.

- Keep dangerous machines well away from the public, and from visitors.

Organizations have a duty to protect non-employees.

- Make sure that control switches are marked so that there's no chance of switching on the wrong machine. If there's a need for an emergency shut-off button, make sure it's clearly labelled, within easy reach, and is coloured red.

Switches should be well marked and easily accessible.

- New machines, hired machines, those being brought back into service after maintenance, and machines standing idle for long periods of time, should be carefully checked before they're used.

Machines which haven't been in use should be checked.

Activity 20

4 mins

What about the machine operator? Try to draw up a brief safety checklist of about **three** or **four** points for operators to carry out whenever they go to use a machine. If it helps, think in terms of a machine used at your own place of work.

_____
_____
_____
_____
_____

A good checklist for an operator might include the following:

- make sure all guards are in place;
- make sure all interlocks are working properly – if necessary, by testing them;
- learn how to **switch off** the machine **before** switching it on;
- put on any specified protective clothing;
- do not wear anything which could possibly get caught up in the machine – loose clothing, chains or rings for instance; long hair should be covered by a hairnet or hat;

Session C

- make sure there's nothing that can get in the way of moving parts, such as loose materials;
- do not use the machine without being trained and authorized;
- make sure there is no sign on the machine that says it is defective or dangerous to use (the supervisor should be told at once if the machine appears not to be working correctly);
- make sure work surfaces and the general area is clean and tidy, and help to keep it that way.

2.1 Dangerous machines

Some machines used in offices, agriculture and industry have special rules. There are certain machines which are 'prescribed dangerous machines', including:

- guillotines;
- mixers;
- portable machinery;
- bacon slicers;
- power-operated wrappers and slicers.

Under the Factories Act, 1961, young people (that is, those 16 or over, and under 18 years) are prohibited from working at any prescribed dangerous machine, unless they have:

- been fully instructed as to the dangers arising in connection with it and the precautions to be observed;
- have received sufficient training in work at the machine;
- are under adequate supervision by a person who has a thorough knowledge and experience of the machine.

Ensure that young persons are trained and supervised.

Of course, all people using such machines should be trained. Young people need special protection because they are the ones most at risk.

Other kinds of machines are covered by Regulations which require certain kinds of guards, to prevent people coming in contact with moving parts. These include:

- abrasive wheels;
- woodworking machines;
- horizontal milling machines;
- power take-off shafts.

Using untrained operators, or unguarded machines, may be against the law.

Session C

3 Preventing falls

It is possible to place falls from accidents into one of three categories:

- people tripping or falling at the same height or level;
- people falling from one height to another;
- something falling onto someone.

3.1 Falls at the same height

We tend to think of falls as being something dramatic, involving falling from a height but falling at the same height as yourself can result in serious injury.

Activity 21

3 mins

Jot down at least **two** typical causes of people having accidents by tripping or slipping at the same level. For example, one cause might be a damaged floor or ground surface.

_____
_____
_____
_____
_____

You may have noted some of the following causes:

- a damaged floor or ground surface, such as an uneven pavement;
- a wet, greasy or icy surface;
- an unevenness in the surface, such as an unexpected step;
- poor lighting;
- some object on the surface;
- unsuitable or worn footwear.

These are all causes owing to the conditions on the floor or ground. You may have also considered possible causes arising from the state or actions of the person who falls, such as being drunk, having poor eyesight or running instead of walking.

Session C

Activity 22

Draw up a checklist to prevent accidents resulting from falling at the same level, based on the list above. One point might be: 'Check surfaces, to ensure they're in good repair.' Try to include at least **five** points in your list.

___
___
___
___
___
___
___
___

A possible checklist is given below. Compare it with yours.

- Don't allow people to walk on a wet floor unless they have suitable footwear.
- Check surfaces, to ensure they're in good repair.
- Put up barriers or portable signs to prevent people walking on temporarily wet and slippery floors.
- Put down salt in icy weather, on outdoor walking surfaces.
- Try to eliminate surface unevenness; if this isn't possible, use signs such as 'Mind the step!'
- Keep areas where people walk well lit.
- Keep surfaces free from clutter.
- Don't allow running or 'larking about'.
- Permit no intoxication, under any circumstances.

3.2 Falls from one level to another

Falls of this kind may be:

- through or into openings: holes, trap-doors and so on;
- through or from roofs;
- from ladders or down stairs;
- while working at heights, such as on a scaffold.

Session C

Activity 23

What do you think are the main reasons that might cause someone to fall into or through an opening such as a hole, pit, trench or trap-door?

___

You could say there were two main reasons:

- an opening not being properly guarded or covered;
- the person being unaware of the opening.

Often, both these reasons apply.

Typically, accidents occur when:

- someone leaves an opening unmarked and unguarded, such as removing a man-hole cover and then leaving it as a hazard for the unwary;
- a cover is inadequate: for example, a trap-door not being capable of supporting a person's weight;
- it is not clear that a board or sheet of metal is covering a hole.

- Two men in a yard, seeing a corrugated sheet lying on the ground, picked it up and started walking, without being aware of the hole it covered. The man behind fell in and was decapitated by the sheet.

The way to prevent such accidents is to ensure that:

- openings are guarded, well marked and well lit;
- covers can withstand any expected load or impact;
- covers are clearly identified.

3.3 Ladder safety

A fall from a ladder may occur when:

- the ladder isn't secured: ladders should preferably be secured by tying at the top, or else at the sides or bottom;
- someone works too high on the ladder without a handhold;
- the ladder is weak or damaged;
- the ladder is at an unstable angle: ladders should have a slope of four units up to one out from the base;
- the ladder is placed against a fragile surface, such as a plastic gutter;
- the ladder is placed on an insecure footing, such as loose flagstones or soft earth;
- the person climbing the ladder is carrying or holding something, having no hands free for holding on to the ladder;
- the ladder is the wrong one for the job – when it is too long or too short, for example.

Another possible hazard is leaving a ladder where a person or a vehicle could collide with it.

3.4 Working on roofs

Every year, more than twenty people die at work by falling through or off roofs.

If it is part of the job of your team to work on roofs, you should have received instruction on the dangers and safeguards.

Generally, accidents are caused when people:

- fall through roofs by walking on fragile materials, wrongly assuming the material will hold their weight;
- fall off roofs by not using suitable crawling boards or proper edge protection;
- injure themselves in other ways, when they ignore hazards such as high winds, or fumes from flue outlets.

To prevent these kinds of accidents, roof workers should:

- never make assumptions about the strength of materials;
- always use proper equipment;
- never ignore hazards.

Isolation barriers should again be used when there is a risk of objects falling on to people below, or where ladders or scaffolding might be walked or driven into.

Session C

3.5 Falling objects

Injuries from falling objects are unfortunately also very common. This fact is recognized in many industries such as building, chemical engineering and so on, and workers in such industries are issued with safety helmets as a matter of course.

Such accidents often come about through errors on the part of people working at high levels or controlling heavy moving equipment. Where activities are taking place above eye level, people on the ground need to be especially alert.

Activity 24

4 mins

Draw up a checklist of **three** or **four** points that might help prevent accidents from falling objects. Consider the area where you work, if this kind of hazard exists there.

_____
_____
_____
_____
_____
_____

Compare your list with this one.

- Don't walk under loads that are suspended, such as from a crane.
- Don't throw anything down from above; get it lowered safely.
- Always wear protective clothing such as helmets and safety shoes where there is a hazard of falling objects.
- Don't leave tools near to the edge of surfaces, or in an unstable position.
- Stack materials with care, making sure, first that they can be stacked, and second, that they're stable.
- Use isolation barriers where there is a hazard of falling objects.

4 Electrical hazards

Electricity is dangerous. It can cause electric shock, explosion and fire.

And yet nearly everybody, at home or at work, uses electricity and electrical appliances every day of their lives.

Session C

Activity 25

Assuming you are not a trained electrician, what can you do to minimize the dangers from electricity? Try to jot down at least **three** actions or precautions you would take.

___

___

___

___

___

___

Some of the things you may have written down are the following.

- Socket outlets should not be overloaded by using adapters. This is dangerous and could cause a fire. If necessary, a multiplug socket block should be used instead – although even these can of course become overloaded
- No one should be allowed to open covers which give access to live electrical parts, unless they are fully trained and competent to do so.
- Faulty apparatus – or any machine which is suspected as being faulty – should be taken out of service and a do not use label placed on it.
- Everyone who uses a machine should be shown how to isolate the power from it in case of emergency. There should be a well-marked switch or isolator close to every machine.
- All electrical installations to be checked regularly by a trained electrician.
- Workteams should be trained to make sure that power sockets are switched off before plugging tools into them.

4.1 Dealing with electric shock

If training in dealing with electric shock is available to you and your team, it's a good idea to take it. As a first step your company should be able to get a copy of the 'electric shock placard' which tells you what to do. But resuscitating someone takes training and practice.

Session C

4.2 Static electricity

The problem doesn't all come from mains electricity. You've probably noticed that you can get a slight electric shock by touching a metal object like a door handle after walking across a synthetic carpet. This is called **static electricity**. Static electricity isn't always dangerous in itself, **but**:

- the shock from static electricity can cause an involuntary movement which could result in an accident;
- sparks generated by static electricity can be very dangerous near flammable liquids, or organic powders like grain or tea dust; they can cause the vapour or dust to ignite, sometimes with explosive results.

For this second reason, whenever these materials are being transported or stored, earthed metal containers should be used, as plastic containers can cause a static charge to build up. Where there is a hazard from static electricity, precautions can be taken to prevent its build-up – such as earthing. Special anti-static clothing and footwear may be worn. You may need specialist advice on this.

4.3 Working outdoors

Some points to remember when electricity is used outdoors:

Socket outlets used outdoors may need to be of a special design and protected by a residual current circuit-breaker. If electrical tools are used outdoors, it's safest to use low voltage equipment or isolating transformers, with the long lead on the low voltage side.

As with all aspects of electricity:

If you aren't sure – get advice about it!

- Working under or near overhead power lines is very dangerous. All machines should be kept at least 15 metres away, or flashover may occur, in which the high voltage strikes the machine. Before working near overhead power lines, talk to the local electricity company.
- If you have to dig holes in the ground, you should always assume that there are buried live electricity cables. You can use cable-detecting equipment to confirm this. Again, you should talk to the electricity company, and use cable avoidance tools. Don't forget that other cables and pipelines may also be present.

Session C

5 Maintenance work

As we've already discussed, over a quarter of fatal accidents in industry occur during, or as a result of, maintenance. There are special problems related to maintenance work.

Activity 26

3 mins

Maintenance work is often done outside the normal working hours of a company, and also may be out of sight of the day staff. A good example of this is in a retail store, where maintenance is usually carried out screened off from the public and staff, and often when the store is closed. Make a note of **three** or **four** problems that might arise in this situation, and which may make the work of maintenance hazardous.

_____

_____

_____

_____

Some problems you may have jotted down are as follows. All of these difficulties may arise in any maintenance situation.

- Normal safeguards may be deliberately bypassed during the maintenance work, and other safeguards not put in their place. For example, electrical wiring may be left uncovered, or handrails removed.
- Maintenance work may have to be carried out in cramped, poorly lit and unusual conditions. Also, where space is limited, adequate supervision may be impractical.
- Work may not always be properly completed. This might result, for example, in accidents to sales staff switching on equipment following the maintenance work.
- Management will be trained to think about the safety of the customers in the store, but may not pay so much attention to safety of maintenance personnel. This may be especially true when the shop is closed, and there are no customers to worry about.
- There may be insufficient communication between operating staff and maintenance workers, to indicate the condition of equipment before maintenance work is commenced.
- Outside contractors may often be employed to carry out maintenance; these people may perform very specialized work. Management and other staff may assume that 'they must know what they're doing' and abdicate any responsibility for the safety of the contractor staff.

52

Session C

Front line managers can play an important role in reducing the number and severity of accidents during maintenance.

Activity 27

In a typical situation, a manager is in charge of some plant or machinery that requires maintenance and/or repair. For this it is necessary to call in other people, from inside or outside the organization.

With the points listed above in mind, draw up a brief checklist, of **three** or **four** key points, for a manager in these circumstances.

_____

_____

_____

_____

Such a checklist may include the following points.

> A permit to work system is a management tool designed for high risk tasks, to ensure that a safe system is in place and is adequately implemented.

- Have you explained precisely what the condition of the plant or machinery is, as far as you are able to determine?
- Have you told the maintenance staff how to isolate the plant or machinery?
- Have you provided any appropriate safety equipment, including first aid materials, and explained how emergency help can be obtained?
- After the work is completed, have you discussed the condition of the equipment with the maintenance staff, before attempting to put it back into service?
- Is a permit to work required? For example, a permit to work is essential where excavations are carried out on industrial premises.

6 Manual handling

More than a quarter of the accidents reported each year are associated with manual handling – the transporting or supporting of loads by hand or by bodily force. Back strain, sprains and limb breakages are relatively common.

If you ask someone to lift up a heavy object and carry it across the floor, he or she may go about the task in the wrong way, and finish up hurting themselves. And yet the correct techniques are not difficult to learn.

Before any manual handling is done, you will need to stand back and think about the task, and the possible alternatives. The following flowchart shows the steps that should be taken whenever you decide to have something moved manually.

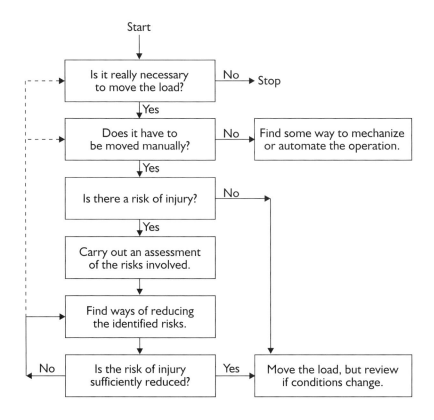

Again, you will notice, the subject of risk assessment has arisen. We'll look at this again shortly.

6.1 Handling techniques

Once you are sure that manual handling is necessary, proper techniques must be applied. These are not easily learned, and training should be given.

The following is an example of good handling technique for picking up a box and carrying to a table or bench.

First, stop and think about how you will tackle the job. You need to plan the lift, and ask yourself:

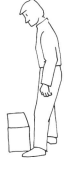

- where you intend to place the load;
- whether you should get someone to help you;
- if you have access to handling aids (such as trolleys) that would be useful;
- whether there are any obstructions (such as packing materials) that should be moved first;
- whether there is a table or bench that you could use to rest the load on, say mid-way between floor and shoulder height.

Session C

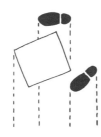

Use your feet as a balanced and stable base for lifting. Make sure your footwear is appropriate, and that your clothes do not impede your movement. Stand feet well apart, with your leading leg slightly forward.

> **EXTENSION 7**
> You can find illustrations and descriptions of safe techniques in the HSE booklet *Manual Handling – Guidance on Regulations*, which also gives useful guidance on the law. The ones here have been adapted from this booklet.

Take up a good posture. Bend your knees but do not overflex them. Keep your back straight, with your chin tucked in. You may need to lean forward a little over the load to get a good grip. Your shoulders should be level and should face in the same direction as your hips.

Grip the load firmly so that it feels secure. Your arms should be kept within the boundary formed by your legs. Bend your fingers round the load if possible. You may find that you have to change your grip during the lift, and if so you should make the changes as smoothly as possible. The heaviest side of the load should be held next to you.

Lift smoothly. Keep control of the load, and don't make sudden movements.

Avoid twisting the trunk when you move to the side – move your feet instead.

Keep the load close to your body. Keep it close to the body for as long as possible.

Rest the load before manoeuvring it into place. It is usually easier to slide a load into position, rather than twisting and moving your body while holding the load.

Session C

6.2 Risk assessment of manual handling

On the next two pages, you can see an example of a detailed checklist that might be used when carrying out such an assessment.

Checklist forms have been taken from the HSE booklet *Manual Handling – Guidance on Regulations*.

 Activity 28

This Activity may provide the basis of appropriate evidence for your S/NVQ portfolio.

The checklist is taken from HSE booklet *Manual handling – Guidance on Regulations*, and you will find it useful to have this booklet with you when you perform the Activity.

If manual handling is carried out in your workplace, use the checklist to assess a particular operation. Fill out all parts of the form, and try to come up with an overall assessment of the operation, and suggestions (if appropriate) for remedial action.

If there is no suitable operation for you to assess, you could go through a mock exercise of (say) planning the movement of a heavy box to an awkward location. Be sure to follow the guidelines for manual handling, which are more fully explained in the HSE booklet.

By all means involve your team members in this Activity, and get help if you can (perhaps from a safety officer, or your manager). You can photocopy the form if you want to, either from this workbook or from the HSE booklet.

Session C

Manual handling of loads
ASSESSMENT CHECKLIST

SUMMARY OF ASSESSMENT	Overall priority for remedial action: Nil / Low / Med / High (Circle as appropriate)
Operations covered by this assessment: _____ _____ _____	Remedial action to be taken: _____ _____ _____
Locations: _____	Date by which action is to be taken: _____
Personnel involved: _____ _____	Date for reassessment: _____
	Assessor's name: _____
Date of assessment: _____	Signature: _____

SECTION A Preliminary

Q1 Do the operations involve a significant risk of injury?   [YES] [NO]

If YES, go to Q2. If NO, the assessment need go no further. (If in doubt, answer YES.)

Q2 Can the operations be avoided / mechanized / automated at reasonable cost?   [YES] [NO]

If NO, go to Q3. If YES, proceed and then check that the result is satisfactory.

Q3 Are the operations clearly within the guidelines?   [YES] [NO]

If NO, go to Section B. If YES you may go straight to Section C if you wish.

SECTION B Detailed assessment – see next sheet.

SECTION C Overall assessment of risk

Q What is your overall assessment of the risk of injury? Insignificant / Low / Medium / High

If not 'Insignificant' go to Section D. If 'Insignificant', the assessment need go no further.

SECTION D Remedial Action

Q What remedial steps should be taken, in order of priority?

a _____

b _____

c _____

d _____

e _____

AND FINALLY:

- Complete the SUMMARY above.
- Compare it with your other manual handling assessments.
- Decide your priorities for action.

TAKE ACTION – AND CHECK THAT IT HAS THE DESIRED EFFECT.

Session C

Questions to consider: (If the answer to a question is 'Yes' place a tick against it and then consider the level of risk)	Level of risk (Tick as appropriate)				Possible remedial action (Make rough notes in this column in preparation for completing Section D)
	Yes	Low	Med	High	
The tasks – do they involve: - holding loads away from trunk? - twisting? - stooping? - reaching upwards? - large vertical movement? - long carrying distances? - strenuous pushing or pulling? - unpredictable movement of loads? - repetitive handling? - insufficient rest or recovery? - a workrate imposed by a process?					
The loads – are they: - heavy? - bulky/unwieldy? - difficult to grasp? - unstable/unpredictable? - intrinsically harmful (e.g. sharp/hot?)					
The working environment – are there: - constraints on posture? - poor floors? - variations in levels? - hot/cold/humid conditions? - strong air movements? - poor lighting conditions?					
Individual capability – does the job: - require unusual capability? - hazard those with a health problem? - hazard those who are pregnant? - call for special information/training?					
Other factors – Is movement or posture hindered by clothing or personal protective equipment?					

When you have completed Section B go to Section C.

Session C

7 Fire hazards

Fires are easily started and often difficult to control.

Your organization will by law have to provide fire extinguishers and issue instruction about what to do in case of fire.

So what more can you and your team do to reduce fire hazards and risks?

Activity 29

Jot down some steps you think that someone in your position could take to cut down fire hazards, and the risks from fire. Try to list **four** actions. Think in terms of your own place of work.

_____
_____
_____
_____
_____
_____
_____

The points you have listed may include some of the following. You could:

- try to minimize the quantities of flammable materials that are kept in the workplace – what is kept must be stored safely;
- make sure that staff don't create unnecessary risks by leaving flammable items about, like oil-soaked rags;
- make sure that 'No smoking' notices are obeyed;
- keep working areas and machines clean; for example, grease should be removed frequently from ducts and cooker extractors;
- make sure that, if rubbish has to be burned, it is done well away from buildings, and that fires are kept under control; extinguishers should be kept handy;
- make sure all staff know how to raise the fire alarm, and that at least some of them know how to use the extinguishers;
- ensure that no one blocks fire exits;
- never let anyone prop open fire doors.

If you need more information or advice about particular hazards, talk to the people in your organization. Failing this, contact HSE's Information Centre at Broad Lane, Sheffield S3 7HQ. Tel: 01742 892345

If you use flammable liquids, gas cylinders, oxygen or other materials that present special fire risks, then you should be aware of the extra hazards, **and** know how to deal with them.

59

Session C

8 Protective equipment

Hundreds of accidents are made worse every year through workers not using adequate protective clothing and other equipment, often referred to as PPE (personal protective equipment).

Protective equipment is designed only as **a last line of defence** against hazards. Nevertheless, it can often save people from death or serious injury.

> **EXTENSION 3**
> This book is listed on page 94. This quote is taken from page 7.

But in spite of this fact, people don't always use the equipment they're provided with. As Jeremy Stranks says in his book: *Management Systems for Safety*

'It is by no means a perfect form of protection in that it requires the person at risk to use or wear the equipment all the time they are exposed to a particular hazard. People simply will not always do this!'

Activity 30

3 mins

Many serious accidents occur every year in cases in which employees have been provided with the proper protective clothing, and have been told of the need for it, but don't use it.

Can you think of any reasons why employees may fail to use protective clothing, even though the clothing is available and they're aware that it should be worn? If protective clothing is used in your workplace, you may be able to draw on your own experience to answer this question.

_____

_____

_____

_____

Some possible reasons are that they:

- feel it's 'too much trouble' to put the clothing on;
- don't really believe any hazard exists;
- may lose time (and pay) by stopping to put the clothing on;
- find the clothing awkward to use, or uncomfortable;
- know the clothing provided is defective in some way;
- believe it isn't necessary for 'just a five minute job'.

Session C

Because protective equipment is often inconvenient, it's important to be vigilant against any tendency to do without it.

Activity 31

3 mins

What steps can the first line manager take to encourage the use of proper protective clothing and equipment at all times? What steps could **you** take? Try to list **four** steps.

_____
_____
_____
_____
_____

A manager or team leader can and should:

- **identify the circumstances** when the protective equipment should be used;
- **assess** the PPE, to make certain it is suitable for the risk;
- make sure that the leader and the team are **fully trained** in its use and aware of its importance;
- make sure the protective equipment is **available**, and in good order – which means that it has to be properly **maintained**;
- make adequate provisions for the **storage** of PPE;
- ensure that every user knows **why** the equipment is necessary, and what its **limitations** are – this means, **instruction**, **information**, and **training**;
- **insist** on the use of the protective equipment and of the workteam developing the **habit** of using it;

If protective clothing can prevent injury or illness, it should not be optional.

- put up safety signs to **remind** people about specific protective equipment;

- set a good **example** by always wearing the proper equipment when it is appropriate to do so;
- **checking regularly** that PPE is used, and doing spot checks if necessary.

Managers and team leaders can only do so much, however. Sooner or later individuals have to take on responsibility for themselves.

Some of the various types of protection are:

Session C

Part of body	Typical hazards	Typical equipment	Comment
Eyes	<ul><li>flying particles;</li><li>dust;</li><li>chemical splashing;</li><li>flying molten metal;</li><li>vapours and gases;</li><li>infectious materials;</li><li>radiation.</li></ul>	Spectacles, goggles, helmets and face screens.	Eye protection to an approved specification must be used for some processes. You should make sure you know whether this is the case with the processes you and your staff work with.
Hearing	<ul><li>Noise</li></ul>	Muffs or earplugs – but choose with care, as many are ineffective.	Hearing is susceptible to many kinds of noise. Extremely loud noise is dangerous even for brief exposures.
Head and neck	<ul><li>falling objects;</li><li>hair entanglement;</li><li>chemical splashing;</li><li>working in extreme climates or temperatures.</li></ul>	Hats, caps, helmets, hairnets, hoods and skull-caps.	As with all equipment, the type of headgear must be effective against the known hazards.
Hands and arms	<ul><li>cuts by anything sharp – even by a sheet of paper in an office;</li><li>burns;</li><li>crushing;</li><li>contamination by infectious materials;</li><li>splashing with chemicals;</li><li>electric shocks;</li><li>being hit by moving objects;</li><li>abrasions.</li></ul>	Gloves, mitts, armlets, chainmail gloves.	Gloves intended for protection against sharp objects may be very poor protection against some chemicals. But if gloves reduce the ability to grip, or if the edges of the gloves might catch in a machine, it could be safer not to wear them.
Feet and legs	<ul><li>slipping;</li><li>cuts and abrasions;</li><li>punctures (e.g. stepping on a nail);</li><li>falling objects;</li><li>heavy pressures;</li><li>chemical and metal splashing;</li><li>wet surfaces;</li><li>electrostatic build-up.</li></ul>	Safety boots with steel toe-caps, reinforced shoes, gaiters and leggings, knee-pads.	
Body	<ul><li>heat, cold and bad climatic conditions;</li><li>chemical and metal splashing;</li><li>spray from pressure leaks or spray guns;</li><li>impact.</li></ul>	Coats, aprons, jackets, boiler suits, high visibility clothing.	Materials may be anti-static, non-flammable, chain mail, chemically impermeable and insulated (against heat or cold), and so on.

Session C

9 Day-to-day tasks

As a manager, you have a responsibility for the general condition of your immediate working environment.

9.1 Housekeeping

In any workplace, housekeeping – that is, general organization, cleanliness, tidiness and maintenance – plays an important role in helping to prevent accidents.

Activity 32

3 mins

Give an example of an accident that might be prevented through good housekeeping.

_____

_____

_____

Some of the hazards that may lurk in any workplace, which can be easily be spotted, and can often be eliminated by simple housekeeping, are:

- articles left on the floor, which may trip the unwary;
- slippery floors, as a result of being wet, or incorrectly polished;
- clutter, which makes people bump into things or fall over;
- sharp corners on furniture, which might injure passers-by;
- items stored at above head height, with no proper provision for getting them down without over-stretching;
- tools left lying about that may be sharp or otherwise dangerous;
- blocked gangways, or a badly designed layout, leading people into hazards in their efforts to get past.

These are just some of the more common hazards. You may have thought of other ways that accidents might be caused.

Apart from helping to eradicate particular hazards, good housekeeping can have a positive effect on the quality and standards of work and behaviour. When employees see high standards being set, they tend to follow them.

If you work in a shop or some other place used by members of the public, then it is even more important to take care of the premises.

Session C

9.2 Stress

But there are other factors, besides housekeeping, which can increase the risk of accidents in the local environment. These include:

- poor lighting;
- high levels of noise;
- too much or too little heat;
- inadequate ventilation;
- the effects of chemicals.

Of course, besides being unpleasant and unhealthy, most of these can be hazardous in themselves. Sustained high noise levels, for example, can cause ear damage, and people die from lack of air. But even when they have no immediate or direct effects, these factors often result in **stress**.

It's a known fact that stress helps to cause accidents. Even when they are simply uncomfortable, through having to wear unsuitable clothing for example, people are more likely to lose their temper and be irritable. This irritability can result in a lack of control over movement and action, which in turn can increase the likelihood of an accident.

Activity 33

Apart from their physical environment, can you think of **one** other factor which may result in increased stress in people at work? Thinking about your own workplace may help you answer this.

_____

_____

You may have answered:

- too much work;
- difficult relationships with colleagues or superiors;
- lack of training;
- lack of information.

There are any number of things that put stress on people.

As a manager or team leader, you will need to take account of causes and symptoms of stress, if you want to reduce the risk of accidents.

Anyone under stress is more prone to have an accident.

Session C

One very important cause, and result, of stress is **fatigue**. That's why rest and meal breaks are important.

To sum up this section, the lessons to be learned are that:

- working areas should be kept clean, tidy and well maintained;
- good housekeeping prevents accidents;
- discomfort and fatigue increased stress, and stress causes accidents;
- signs of stress among your team should be taken seriously.

Activity 34

Portfolio of evidence A1.2

This Activity may provide the basis of appropriate evidence for your S/NVQ portfolio. If you are intending to take this course of action, it might be better to write your answers on separate sheets of paper.

Look back through this session, and select one of the areas covered that is applicable to your own situation: machinery safety; preventing falls; electrical hazards; maintenance; fire hazards; protective equipment; or day-to-day hazards. (You should already have completed Activity 28 on manual handling.)

Undertake a thorough review of the accident prevention measures currently in place in your work area, in respect of your chosen topic. You can use the points listed in this session as a checklist, or you may want to make up your own checklist, based on experience or on another document.

Get your team involved in this exercise if possible, and don't be afraid to ask for help from your colleagues.

Self-assessment 3

1 The following statements have been split in half. Match the correct second half with each first half.

a	Accident prevention shouldn't depend on	i	just good practice – it's the law.
b	Ensure that young persons are	ii	people always obeying the rules.
c	Inspection and maintenance of	iii	unguarded machines, may be against the law.
d	It's useful to think about	iv	equipment are essential.

Session C

e	Keep untrained and	v	well marked and easily accessible.
f	Keeping the guards in place is not	vi	protect non-employees.
g	Machines which haven't been in use	vii	unauthorized people away from dangerous machines.
h	Organizations have a duty to	viii	should be checked.
i	Switches should be	ix	trained and supervised.
j	Using untrained operators, or	x	what might go wrong.

2 What are the **two** most important points to bear in mind about good housekeeping?

_____
_____
_____
_____

3 List **three** ways of reducing the risk from electric shock.

_____
_____
_____
_____

4 What are the **three** kinds of falls that most commonly occur in workplaces?

_____
_____
_____
_____

5 List **four** precautions you might take to help prevent a ladder accident.

_____
_____
_____
_____

The answers to these questions can be found on pages 97–8.

Session C

10 Summary

- In accident prevention, it's useful to think about what might go wrong.

- Accident prevention shouldn't depend on people always obeying the rules.

- When dealing with machine safety:
  - keeping the guards in place is not just good practice – it's the law;
  - inspection and maintenance of equipment are essential;
  - keep untrained and unauthorized people away;
  - switches should be well marked and easily accessible;
  - machines which haven't been in use should be checked;
  - ensure that young persons are trained and supervised: using untrained operators, or unguarded machines, may be against the law.

- To prevent falls at the same level, some of the points mentioned were to:
  - check surfaces, to ensure they're in good repair;
  - put up barriers to prevent people walking on temporarily wet and slippery floors;
  - keep areas where people walk well lit;
  - keep surfaces free from clutter;
  - don't allow running or 'larking about';
  - permit no intoxication, under any circumstances.

- To prevent people falling from one level to another, ensure that:
  - openings are guarded, well marked and well lit;
  - covers can withstand any expected load or impact;
  - covers are clearly identified.

- To prevent falls from ladders:
  - secure ladders by tying at the top, or else at the sides or bottom;
  - don't allow anyone to work on a ladder without a handhold;
  - check to see that ladders are not weak or damaged;
  - place ladder at a slope of four units up to one out from the base;
  - do not place a ladder against a fragile surface, such as a plastic gutter;
  - make sure the ladder is placed on an secure footing;
  - use the correct ladder for the job.

- Roof workers should:
  - never make assumptions about the strength of materials;
  - always use proper equipment;
  - never ignore hazards.

- To protect against being hurt by falling objects:
  - don't walk under loads which are suspended, such as from a crane;
  - don't throw anything down from above; get it lowered safely;

- always wear protective clothing such as helmets and safety shoes where there is a hazard of falling objects.
- don't leave tools near to the edge of surfaces, or in an unstable position;
- use isolation barriers where there is a hazard of falling objects.

■ For electrical safety:
- no-one should be allowed to open covers which give access to live electrical parts, unless they are fully trained and competent to do so;
- there should be a well-marked switch or isolator close to every machine;
- all electrical installations to be checked regularly by a trained electrician;
- workteams should be trained to make sure that tools and power sockets are switched off before plugging them in.

■ For maintenance safety:
- explain precisely what the condition of the plant or machinery is, as far as you are able to determine;
- tell the maintenance staff how to isolate the plant or machinery;
- provide appropriate safety equipment, including first aid materials, and explain how emergency help can be obtained;
- discuss the condition of the equipment with the maintenance staff, before attempting to put it back into service;
- check whether a permit to work is required.

■ Before any manual handling is done, you will need to stand back and think about the task, and the possible alternatives. Is manual handling really necessary?

■ Training should be given in correct handling techniques.

■ To protect against fire hazards:
- make sure that 'No Smoking' notices are obeyed;
- keep working areas and machines clean; for example, grease should be removed frequently from ducts and cooker extractors;
- make sure that, if rubbish has to be burned, it should be done well away from buildings, and fires kept under control; extinguishers should be kept handy;
- make sure all staff know how to raise the fire alarm, and that at least some of them know how to use the extinguishers;
- ensure that no one blocks fire exits or props open fire doors.

■ With regard to personal protective equipment (PPE):
- identify the circumstances when the PPE should be used;
- assess the PPE, to make certain it is suitable for the risk;
- make sure the team are fully trained in its use and aware of its importance;

Session C

- make sure the PPE is available, and in good order;
- make adequate provisions for the storage;
- provide instruction, information and training;
- insist on the use of the protective equipment and of the workteam developing the habit of using it;
- check regularly that PPE is used, and carry out spot checks if necessary.

■ In any workplace, housekeeping – that is, general organization, cleanliness, tidiness and maintenance – plays an important role in helping to prevent accidents, and can have a positive effect on the quality and standards of work and behaviour.

■ Anyone under stress is more prone to have an accident.

Session D Coping with accidents

1 Introduction

It's a fact of life that accidents do happen, whatever we do to try to prevent them. When they occur, there is more work for the first line manager to do.

For one thing, you need to be prepared for the worst, and to train people in dealing with emergencies.

Then, after it's over, an accident has to be reported, and perhaps investigated.

2 Dealing with accidents and abnormal occurrences

It may well be part of your job to handle things in an emergency at work. How would you go about it?

Let's first list some of the accidents and emergencies which might occur in a place of work.

Session D

Activity 35

Think about your own place of work. Jot down **two** or **three** examples of any accidents which you know have occurred in the past.

_____
_____
_____
_____
_____
_____

Now try to think of situations where an accident came close to occurring – perhaps someone fell but wasn't injured, or someone didn't bother to wear protective clothing when they should have, but 'got away with it'.

_____
_____
_____
_____
_____

Depending on the type of activity which takes place and the circumstances, some or all of the following are possible:

- a fire;
- an explosion;
- electric shock;
- someone falling, or something falling on someone;
- a flood;
- a release of dangerous gases;
- a poisoning;
- a release of radioactivity;
- someone being hurt or trapped by a machine;
- someone getting burned;
- a road or other transport accident;

You may have thought of (or have heard about) other kinds of accidents. The list of things that might possibly go wrong is bound to be a long one.

Your company may well have written instructions as to what people should do in the event of an accident. There will almost certainly be fire instructions, and there should be instructions for dealing with accidents arising from any special risks involved in the different areas of work.

Session D

Activity 36

Instructions which tell people what to do in the event of an emergency would normally include, for example:

- saying what might happen, and how an alarm would be raised.

Try to locate any emergency instructions that are displayed at your place of work, and, having read the wording carefully, try to list two further points that emergency instructions might include.

_____
_____
_____
_____
_____
_____

You may have listed any of the following:

- how to call the emergency services;
- how to reach safety;
- who should be informed of the situation;
- where to find rescue equipment;
- saying what might happen, and how an alarm would be raised;
- the names or titles of people who would take charge in the event of an incident;
- where to find first aid equipment, and the names of people who are trained in giving first aid;
- how to shut down plant or make equipment safe.

First line managers are expected to take a lead in implementing safety arrangements. You may want to give some thought to whether your team are sufficiently well informed about how to deal with emergencies.

Now let's look at what should be done when an emergency does occur.

In the following activities, imagine that you arrive at the scene **immediately** after each incident has happened.

Session D

Activity 37

3 mins

Situation 1

- Two workmen are painting at the top of some scaffolding. The scaffolding suddenly collapses and the men are thrown to the ground. They appear to be seriously injured.

You arrive at the scene. What do you do?

The first thing you would have to do is to assess the situation. An important step is to establish whether further harm or damage could yet occur. For example, part of the scaffolding may still be liable to fall. It may not be easy to decide whether the area is safe.

- The most sensible first step in a situation like this is therefore to make sure no one else could get hurt, including yourself. You may need to clear the area.
- You may need to organize other potential helpers in the vicinity, and you would certainly need to call for medical help, perhaps by telling someone to telephone for an ambulance.
- First aid must be given to the injured as soon as possible.

This was an example of a serious accident, and one which would have to be reported, both within the organization, and to the HSE. We will discuss reporting shortly.

Session D

Activity 38

Situation 2

- A tank containing poisonous and flammable liquid ruptures and begins spraying the surrounding area, soaking the clothes of two operators nearby.

You arrive at the scene and assess what has happened. What do you do?

_____

The liquid may well be giving off toxic or asphyxiating fumes, and may be absorbed into clothing. There is also a risk of fire, so this is a very hazardous situation to deal with. A difficult rescue operation may have to be mounted.

- Your first thought, again, must be to prevent the accident becoming worse than it already is. In a number of accidents of this kind, would-be rescuers have become additional victims, as in this example.

- A road tanker was carrying a load of oleum (fuming sulphuric acid) along the M6 motorway. The weather was foggy and, as the driver was taking avoiding action, he swerved and struck the nearside of a stationary container lorry. The collision was sufficiently violent to rupture the tank and some 13 tons of oleum was spilt on to the road. A woman driving a following car stopped near the crash and got out of her car. She slipped and fell into a pool of acid and was burned so severely that she subsequently died; twenty other people also suffered acid burns in the incident.[6]

- Getting expert medical and fire service help is essential, and you will need to describe what has happened. The emergency services – the fire and ambulance crews – will want to know what the liquid is. (This information will be displayed on the tank in the form of a code, whether a road tanker or a fixed tank is involved.)

- As soon as possible, the affected operators must be taken to a safe area and given medical help. Removing clothing may be a priority.

Again, this accident would have to be reported.

[6] Text from Health and Safety Commission Newsletter, April 1987

Session D

Activity 39

Situation 3

- Through faulty electrical wiring, a power tool becomes 'live', giving a severe electric shock to the man using it. The tool then falls onto a metal bench.

You arrive at the scene. What do you do?

_____
_____
_____
_____
_____
_____

The same general approach should be applied.

1 Assess the situation.

2 Make the area safe: in this case the first thing to be done is to switch off the power.

3 Get someone to call expert help: urgent medical help would be needed.

4 Apply first aid to the injured. In a case like this, every second will count.

Unless you are trained in dealing with the kind of emergency described in each incident above, you may feel that you couldn't cope. Although you may not be confident in knowing exactly what to do in a particular situation, there are rules which apply to **any** emergency.

Activity 40

Based on the answers to the three incidents above, summarize what should be done in the event of any emergency, by listing **three** or **four** points.

_____
_____
_____
_____

Session D

The actions you should have listed, in your own words, are:

- **Assess the situation**

  to see what actions are needed.

- **Make the area safe**

  so that no one else is in danger. If this isn't possible, clear everyone away from the area. Make sure there will be no additional victims.

- **Get help**

  from any appropriate source: colleagues, the Ambulance Service, the Fire Service, and/or the Police.

- **Give first aid**

  to any people injured. If necessary, call for medical assistance.

Once the emergency is under control, there are still more actions to be taken.

Activity 41

What further steps should be taken after an accident, once the immediate emergency has been dealt with?

_____
_____
_____
_____
_____
_____

If the accident is to be investigated, **nothing should be moved unnecessarily**. For example, in the case of the accident with the scaffolding, you shouldn't allow anyone to start dismantling the structure, beyond making it safe. Doing this could destroy evidence about the cause of the accident.

With any accident, the names and addresses of any witnesses should be taken.

Also, **the accident has to be reported**. We'll look at the matter of reporting next.

Session D

3 Reporting accidents

Some accidents must be reported as a legal requirement.

Unfortunately, accidents don't always get reported. If it were possible to ask people involved in accidents why they didn't report them, you would probably get answers like:

- 'I didn't think it was serious enough. No-one was seriously hurt.'
- 'Last time this happened I filled out the accident form as I was told. But nothing happened – no one took any notice.'
- 'I haven't got time to sit around filling out forms. I've got important work to do.'

Activity 42

3 mins

Why do you think it is so important to report accidents? Try to give **two** reasons.

_____
_____
_____
_____

Some reasons are as follows.

- To help to prevent further accidents.

    When an accident is reported within an organization, an investigation can be mounted, which may well result in new safety measures being introduced to prevent recurrences.

- To enable compensation claims to be made.

    If an accident is not reported as having occurred at work as soon as it happens, it may be very difficult to prove at a later stage that it actually did, especially if the effects of the accident don't become obvious for some time afterwards. Furthermore, when a claim for compensation is made, it is vital that it be established that the injury occurred at work.

- So that special precautions can be taken.

    A survey of reported accidents may well reveal unsuspected risks or adverse conditions, and so on. The organization can then concentrate on how to prevent related accidents from occurring in the future.

Session D

> POINT OF LAW
>
> The Reporting of Injuries, Diseases and Dangerous Occurrences (RIDDOR) Regulations require organizations to notify the relevant enforcement authority (the HSE) 'by the quickest practicable means' – usually by telephone. Subsequently a report must be made within ten days on the approved form in the event of:
>
> - the **death** of any person as a result of an accident arising out of or in connection with work;
> - any person at work suffering a **specified major injury** as a result of an accident arising out of or in connection with work;
> - any person who is not at work suffering an injury as a result of an accident arising out of or in connection with work, and where that person is taken from the site of the accident to a hospital for treatment in respect of that injury;
> - any person who is not at work suffering a major injury as a result of an accident arising out of or in connection with **work at a hospital**;
> - where there is a **dangerous occurrence**.
>
> A report must also be made as soon as practicable, and in any event within ten days of the accident, where anyone is **incapacitated for work** that he or she might reasonably be expected to do, **for more than three consecutive days**. The report has to be made even if the injured person is not away from work but, for example, performing light duties.

It is part of a manager's job to make sure **all** accidents get reported. Other staff may look to the supervisor, too, to do everything possible to make sure the same accident can't happen again.

Filling out an accident form and then locking it away in a filing cabinet and forgetting about it will do no one any good at all.

4 Investigating an accident

As part of a sound system of safety, organizations should investigate the direct and indirect causes behind accidents at work.

The investigators would want to know a number of things, including:

- what type of accident it was, such as a maintenance accident, a trip or a fall and so on;
- what injuries, if any, were inflicted;
- whether the law was broken;
- whether the accident is a notifiable one;
- if an insurance claim should be made.

Session D

Investigations should begin as soon as practicable, following the occurrence. The more time that passes, the more likelihood there is of the people involved forgetting the details, or of physical evidence being destroyed.

4.1 Who investigates?

Who should take part in running an investigation?

Activity 43

Who would you expect to participate in managing an accident investigation in your organization?

_____

_____

_____

Usually, the question regarding which people take part will depend on the seriousness of the incident. You may be asked to run an investigation, if something happens in your own work area. Alternatively, a more senior colleague, or a safety officer, may take charge. Trade union representatives are normally also invited to participate.

If the accident is a serious one, the HSE may decide to carry out their own investigation. Lawyers representing injured persons, and/or insurance assessors, may also need to be involved.

4.2 What do they do?

The following procedure covers most points of an investigation.

- **Find out the facts**, regarding:
    - the sequence of events that lead up to the accident;
    - the system of work in operation;
    - environmental factors;
    - the plant and equipment involved;
    - the people who were present.

- **Take photographs**, make sketches, and take measurements of the scene and the relevant features. Later, scaled drawings may have to be produced.

Session D

- **Obtain statements**, as soon as possible, from all persons who were involved in, or who observed, the accident. Write down their names and addresses, who they work for, and the reasons they were on the site. For a serious occurrence, interviews should ideally be recorded, and/or take place in front of a third party.

- **Review the facts**, in the light of what has been learned, taking steps to resolve any inconsistencies or conflicting evidence.

- **Get expert help**, if necessary: for example, to examine machinery.

- **Come to a conclusion**, if possible, regarding the causes of the accident.

- **Generate a written report** of the accident, which describes what happened, sets out the causes, and recommends changes to prevent a recurrence.

> It is *not* the purpose of an investigation to find somebody to blame for the accident.

4.3 What happens next?

As a result of the investigation, action needs to be taken, to prevent a further accident. Some of the following questions will have to be answered. Should:

- further investigation take place?
- new instructions be implemented?
- further training take place?
- new systems of work – such as a permit to work system – be installed?
- the job be analysed, to find out whether it should be done in a different way?
- different materials be used?
- different work methods be used?
- the work be done in a different place or environment?
- the work be supervised more closely?
- responsibilities be reassigned?
- expert advice be sought?

Session D

Self-assessment 4

1. What immediate actions should be taken at the scene of any accident?

   a _____

   b _____

   c _____

   d _____

2. List **five** important actions for someone investigating an accident.

   _____

The answers to these questions can be found on page 98.

Session D

5 Summary

- Many kinds of unexpected accidents and incidents may occur, and people in the workplace have to know how to deal with them.

- Some good advice is to make plans on the assumption that what could go wrong, will go wrong.

- When an emergency occurs, the following steps should be followed:
  1 Assess the situation, to see what actions are needed.
  2 Make the area safe, so that no one else is in danger. If this isn't possible, clear everyone away from the area. Make sure there will be no additional victims.
  3 Get help from any appropriate authority: colleagues, the Ambulance Service, the Fire Service, the Police – whichever is appropriate.
  4 Give first aid to any people injured. If necessary, call for medical assistance.

- Accidents must be reported:
  - to help to prevent further accidents;
  - to enable compensation claims to be made;
  - so that special precautions can be taken;
  - for legal reasons.

- Accident investigators should find out:
  - what sequence of events led up to the accident;
  - what system of work was in operation;
  - about any relevant environmental factors;
  - what plant and equipment was involved;
  - who was present.

- The investigators should also:
  - take photographs, make sketches, and take measurements of the scene and the relevant features;
  - obtain statements, as soon as possible, from all persons who were involved in, or who observed, the accident;
  - review the facts, in the light of what has been learned, taking steps to resolve any inconsistencies or conflicting evidence;
  - get expert help, if necessary;
  - come to a conclusion, if possible, regarding the causes of the accident;
  - generate a written report of the accident, which describes what happened, sets out the causes, and recommends changes to prevent a recurrence.

Performance checks

1 Quick quiz

Jot down the answers to the following questions on *Preventing Accidents*.

Question 1 Why is it important to learn lessons from incidents in which nobody gets hurt and no damage is done?

Question 2 What is meant by the statement 'accidents at work are largely caused by poor control and management of safety'?

Question 3 Why is it important for organizations to set safety standards that are measurable, and against which objectives can be compared?

Question 4 What is meant by the expression 'the accident iceberg of costs'?

Question 5 Why must safety strategies in organizations be proactive, rather than reactive?

Question 6 Which are the **three** principal sources of law for health and safety?

Question 7 Briefly summarize an employee's duties under the Health and Safety at Work etc. Act, 1974.

Performance checks

Question 8 Define, in your own words, what risk assessment is.

_____
_____

Question 9 Write down **four** appropriate actions to help prevent accidents on ladders.

_____
_____
_____
_____

Question 10 Why is static electricity sometimes dangerous?

_____
_____

Question 11 Write down **three** questions that you should ask **before** asking people to move heavy loads manually.

_____
_____
_____

Question 12 Note down **four** kinds of hazard that might result in injury to the eyes.

_____
_____
_____
_____

Question 13 Explain briefly why good housekeeping helps accident prevention.

_____
_____

Question 14 What simple advice would you give a manager who is trying to plan for emergencies, and is trying to work out what kind of abnormal occurrences are likely to take place?

_____
_____

Question 15 What are the **four** actions to be performed, in the event of any emergency?

_____
_____
_____

Answers to these questions can be found on page 99–100.

Performance checks

2 Workbook assessment

Read the following case incident and then deal with the instruction that follows. Write your answers on a separate sheet of paper.

- In the HaggarLight Engineering Company, a board meeting was being held. The 'hot' item on the agenda was the company's accident record. In recent months, there had been one serious accident and a couple of 'near misses'.

  The board members were divided over the issue. One argument, put forward by John Polanski the Finance Director, was that it would be wrong to get too excited over the matter. 'OK, we have a little tightening up to do,' John said. 'Some extra training for the fork lift truck drivers, perhaps. This has blemished our otherwise excellent safety record, but there's no need to go overboard. If we put in place the measures that Liam is suggesting, we will no doubt end up paying out a lot of money, and it won't be money spent wisely, in my view.'

  Liam O'Flaherty, the company's works director, had put a set of proposals before the board that went much further than 'a little tightening up'. He considered that a complete review of safety policy and procedures was needed, and he had also put forward a list of immediate actions, including the appointment of a firm of safety consultants to carry out the review. Liam thought that the accident, in which a visiting lorry driver had been run down by a forklift truck, was not a 'one off', but was symptomatic of declining standards, coupled with increasingly cramped working conditions, as the company expanded. Liam's greatest concerns were what he saw as a lack of leadership in safety matters, together with a reluctance, on cost grounds, to put in place even the most basic measures.

  John Polanski continued: 'Apart from making the consultants richer, and us poorer, I can't see where this will get us. Of course, we have to obey the law on safety, but the main aim of the organization is to make a decent profit – we aren't a charity, after all. None of us wants to see anyone getting hurt, but our accident record is no worse than most. You all know how tight money is as a result of our high investment in new machinery over the past two years.'

Put yourself in Liam O'Flaherty's shoes, and put forward a convincing list of points about the importance of implementing effective safety measures. You will need to overcome the board's natural reluctance to spend money unnecessarily.

Make any assumptions about the company that you wish – you may want to use facts from your own organization. However, you should write down what assumptions you have made.

You do not need to write more than a page. Explain each of your points clearly – don't assume your readers will have the same level of understanding of the subject as you.

Performance checks

3 Work-based assignment

Portfolio of evidence A1.2, D1.1, D1.2

 60 mins

The time guide for this assignment gives you an approximate idea of how long it is likely to take you to write up your findings. You will find you need to spend some additional time gathering information, talking to colleagues, and thinking about the assignment.

Your written response to this assignment may form useful evidence for your S/NVQ portfolio. The assignment is designed to help you to demonstrate your personal competence in:

- building teams;
- focusing on results;
- thinking and taking decisions;
- striving for excellence.

What you have to do

Identify a hazard within your own work environment, which might lead to an accident. Describe what this hazard consists of. Then co-operate with your team to carry out a risk assessment, by estimating:

- the likelihood or probability that an accident might occur as a result of the hazard;
- how serious the outcome might be, if it did occur;
- how often the risk is present.

Do not make snap judgements: all these questions require a good deal of thought. You may not feel you have enough experience or knowledge to complete this exercise on your own, so by all means call upon the expertise of colleagues, including any with specific training in risk assessment.

You may decide to calculate a risk rating, using the tables listed in Session B or using a method adopted by your organization. However, the important outcome of this assessment is to determine precisely what actions you recommend taking, in order to eliminate or reduce the risk.

Write your response in the form of a report to your manager.

Reflect and review

1 Reflect and review

Now that you have completed your work on *Preventing Accidents*, let us review the workbook objectives. The first objective was:

- When you have completed this workbook you will be better able to play your part in implementing and maintaining safe systems of work.

A safe system of work was defined as 'the integration of people, machinery and materials in a correct working environment to provide the safest possible working conditions'. Systems are derived from organizational policies, and, to some extent, your scope in implementing safe systems will depend on the structures in place. However, there is plenty to be done at a local level.

- Write down **one** way in which you could improve your team's working conditions, to make them safer.

  _____
  _____
  _____
  _____

- How will you put measures in place to do this?

  _____
  _____
  _____
  _____

- Is your team's working environment safer or more hazardous than it was six months ago? If it is more hazardous, what plans do you intend to make to get them back to the earlier level of safety?

  _____
  _____
  _____
  _____

Reflect and review

The second objective was:

- When you have completed this workbook you will be better able to identify hazards in your workplace, and take effective precautions against them.

We have discussed a number of hazards, including some that helped to cause the seven 'classic' accident types. Hazards exist everywhere: in offices, factories, hospitals, schools, on the road – in all environments that people work. Identifying hazards is a key task of any first line manager, and you will no doubt want to encourage your team in being vigilant in this regard.

- How can you become more systematic in your identification of hazards in your work area?

  _____
  _____
  _____

- How can you improve the precautions already in place?

  _____
  _____
  _____

The third objective was:

- When you have completed this workbook you will be better able to take part in risk assessment.

As a representative of your employer, you have to play your part in assessing the risks to the health and safety of your team to which they are exposed at work, and the risks to the health and safety of visitors and others. We defined risk assessment as:

'an identification of the hazards present in an undertaking and an estimate of the extent of the risks involved, taking into account whatever precautions are already being taken'.

This is an important new responsibility, placed on employers by the law. Every organization must carry out risk assessments. We discussed the fact that three aspects of hazards must be evaluated:

- the likelihood or probability that an accident might occur;
- how serious the outcome might be, if it did occur;
- how often the risk is present.

This obligation means more than carrying out a single exercise. Because new hazards are always appearing, and the level of risk is typically changing frequently, risk assessment must be repeated over and over.

Reflect and review

■ Have all the hazards in your work environment been assessed recently?

_____
_____
_____

■ How can you learn more about the process of risk assessment?

_____
_____
_____

The next objective of the workbook was:

■ When you have completed this workbook you will be better able to identify some important points of health and safety law.

The law on health and safety is embodied in many Acts and Regulations. In this workbook, we have covered only some of them, notably the Health and Safety at Work etc. Act, 1974, and the Management of Health and Safety at Work Regulations, 1992.

■ How can you learn more about the law on health and safety?

_____
_____
_____

The final objective was:

■ When you have completed this workbook you will be better able to cope with, report on and investigate accidents at work.

Session D dealt with these three topics. We noted that there is a procedure that can be applied to all abnormal occurrences and emergencies. The reporting of accidents is important, in order to:

■ help to prevent further accidents;
■ enable compensation claims to be made;
■ allow special precautions to be taken;
■ comply with the law.

Accident investigators want to know:

■ what type of accident it was, such as a maintenance accident, a trip or a fall, and so on;
■ what injuries, if any, were inflicted;
■ whether the law was broken;
■ if an insurance claim should be made.

Reflect and review

- Are you confident that you can handle most emergencies? If not, what do you need to do to become more confident?

  _____
  _____
  _____

- How could you improve your accident reporting procedures?

  _____
  _____
  _____

- How can you find out more about accident investigation?

  _____
  _____
  _____

2 Action plan

Use this plan to further develop for yourself a course of action you want to take. Make a note in the left-hand column of the issues or problems you want to tackle, and then decide what you intend to do, and make a note in Column 2.

The resources you need might include time, materials, information or money. You may need to negotiate for some of them, but they could be something easily acquired, like half an hour of somebody's time, or a chapter of a book. Put whatever you need in Column 3. No plan means anything without a timescale, so put a realistic target completion date in Column 4.

Finally, describe the outcome you want to achieve as a result of this plan, whether it is for your own benefit or advancement, or a more efficient way of doing things.

	1 Issues	2 Action	3 Resources	4 Target completion
Desired outcomes				
Actual outcomes				

Reflect and review

3 Extensions

Extension 1

Book *Classic Accidents*
Author David Farmer
Edition 1989
Publisher Croner Publications Ltd

This small book, besides describing numerous examples of accidents, gives helpful advice on preventing them.

Extension 2

Book *The Costs of Accidents at Work*
Edition 1993
Publisher Health and Safety Executive

Intended to be helpful to business, this book of case studies should help you realize just how costly accidents can be.

Extension 3

Book *Management Systems for Safety*
Author Jeremy Stranks
Edition 1994
Publisher ROSPA/Pitman

If you only want to have one reference book on health and safety, this is as good as any. The publication *Health and Safety at Work* rated it as 'the best book yet on the subject'.

Extension 4

List of principal Acts of Parliament, and Regulations, pertaining to health and safety at work.

The principal Acts:

> Consumer Protection Act, 1987
> Criminal Justice Act, 1991
> Environment Act, 1995
> Environmental Protection Act, 1990
> Factories Act, 1961
> Fire Precautions Act, 1971
> Fire Safety and Safety of Places of Sport Act, 1987
> Health and Safety at Work, etc. Act, 1974
> Mines and Quarries Act, 1954
> Offices, Shops and Railway Premises Act, 1963
> Social Security Act, 1975

Reflect and review

The principal regulations:

>Chemicals (Hazard Information and Packaging for Supply) (CHIP 2) Regulations, 1993
>Construction Regulations, 1961–1966
>Construction (Design and Management) Regulations, 1991
>Construction (Head Protection) Regulations, 1989
>Control of Asbestos at Work Regulations, 1987
>Control of Lead at Work Regulations, 1980
>Control of Industrial Major Accident Hazards Regulations, 1984
>Control of Pesticides Regulations, 1986
>Control of Substances Hazardous to Health (COSHH 2) Regulations, 1994
>Electricity at Work Regulations, 1989
>Gas Safety (Installation and Use) Regulations, 1994
>Health and Safety (Display Screen Equipment) Regulations, 1992
>Health and Safety (Enforcing Authority) Regulations, 1989
>Health and Safety (First Aid) Regulations, 1981
>Health and Safety (Information for Employees) Regulations, 1989
>Highly Flammable Liquids and Liquefied Petroleum Gases Regulations, 1972
>Ionizing Radiations Regulations, 1985
>Lifting Plant and Equipment (Records of Test and Examination, etc.) Regulations, 1992
>Management of Health and Safety at Work Regulations, 1992
>Management of Health and Safety at Work (Amendment) Regulations, 1994
>Manual Handling Operations Regulations, 1992
>Notification of Cooling Towers and Evaporative Condensers Regulations, 1992
>Noise at Work Regulations, 1989
>Personal Protective Equipment at Work Regulations, 1992
>Pressure Systems and Transportable Gas Containers Regulations, 1989
>Provision and Use of Work Equipment Regulations, 1992
>Reporting of Injuries, Diseases and Dangerous Occurrences Regulations (RIDDOR), 1995
>Safety Representatives and Safety Committees Regulations, 1977
>Safety Signs Regulations, 1980
>Simple Pressure Vessels (Safety) Regulations, 1991
>Social Security (Industrial Injuries) (Prescribed Diseases) Regulations, 1985
>Workplace (Health, Safety and Welfare) Regulations, 1992

Extension 5

Book *Safety Representatives and Safety Committees*
Edition 1988
Publisher Health and Safety Commission

Contains Code of Practice, Regulation, and Guidance notes.

Reflect and review

Extension 6 Book   *Essentials of Health and Safety at Work*
                Edition   1994
                Publisher Health and Safety Executive

Extension 7 Book   *Manual Handling (Manual Handling Operations Regulations 1992 – Guidance on Regulations)*
                Edition   1992
                Publisher Health and Safety Executive

Contains just about everything you might want to know about manual handling, including how to comply with the law.

Many of these extensions can be taken up via your NEBS Management Centre. They will either have them or will arrange that you have access to them. However, it may be more convenient to check out the materials with your personnel or training people at work – they may well give you access. There are other good reasons for approaching your own people; for example, they will become aware of your interest and you can involve them in your development.

4 Answers to self-assessment questions

Self-assessment 1 on page 15

1 The correct matches are as follows:

Accident   Any undesired circumstances which give rise to ill health or injury; damage to property, plant, products or the environment; production losses or increased liabilities.

Hazard   The potential to cause harm, including ill health and injury; damage to property, plant, products or the environment; production losses or increased liabilities.

Risk   The likelihood that a specified undesired event will occur due to the realization of a hazard by, or during, work activities; or by the products and services created by work activities.

Danger   An unacceptable level of risk.

Safety   The result of the activities we carry out to keep something or somebody from harm.

2 a The seven 'classic' accident types [iii] keep on happening to different people, in different places.
   b Accidents at work are largely caused by [ii] safety systems out of control.
   c An organization's health and safety policy statement [i] is the starting point for all accident prevention and health promotion.

Reflect and review

Self-assessment 2 on page 37

1 The correct statements are:

   b Proactive safety strategies, rather than reactive ones, are best.
   d Safe processes, safe premises, and safe materials, are all necessary and important in making workplaces safe.

2   a A safe SYSTEM of work is the INTEGRATION of people, machinery and materials in a correct working ENVIRONMENT to provide the safest possible working CONDITIONS.
   b Under HSWA, EMPLOYERS must have regard for the safety of NON-EMPLOYEES who may be affected by the activities of their companies.
   c All EMPLOYEES have duties: to take REASONABLE care for their own safety and that of others; to CO-OPERATE with their employers in matters of safety; not to INTERFERE with or misuse anything provided for safety.
   d Every EMPLOYER must make an ASSESSMENT of the health and safety RISKS of work activities to EMPLOYEES and anyone else who may be affected, and record the findings.
   e MHSWR requires employers to appoint COMPETENT people.

3 Risk ratings are assigned to hazards in order to compare the risks associated with various work operations, and to help decide on the urgency of remedial action required.

4   1 Eliminate the risk.
   2 Enclose the risk.
   3 Install a safety device.
   4 Implement a system of work that reduces the risk.
   5 Provide specific written safety instructions.
   6 Supervise those at risk from the hazard.
   7 Provide training.
   8 Provide general information about safety.
   9 Provide personal protective equipment.

Self-assessment 3 on page 65

1   a Accident prevention shouldn't depend on people always obeying the rules.
   b Ensure that young persons are trained and supervised.
   c Inspection and maintenance of equipment are essential.
   d It's useful to think about what might go wrong.
   e Keep untrained and unauthorized people away from dangerous machines.
   f Keeping the guards in place is not just good practice – it's the law.
   g Machines which haven't been in use should be checked.
   h Organizations have a duty to protect non-employees.
   i Switches should be well marked and easily accessible.
   j Using untrained operators, or unguarded machines, may be against the law.

2 The two most important points to bear in mind about good housekeeping are that it plays an important role in helping to prevent accidents, and can have a positive effect on the quality and standards of work and behaviour.

Reflect and review

3 Some of the ways of reducing the risk from electric shock are the following.

- Don't leave covers open, or allow untrained or incompetent people to open covers over live electrical parts.
- Disconnect faulty apparatus and place a do not use label on it.
- Show all users of electrical machines how to isolate the power from it in case of emergency.
- Place a well-marked switch or isolator close to every machine.
- Have all electrical installations checked regularly by a trained electrician.

4 Falls that most commonly occur in workplaces include:

- people tripping or falling at the same height or level;
- people falling from one height to another;
- something falling on to someone

5 To help prevent ladder accidents, you could:

- make sure ladders are secured by tying at the top, or else at the sides or bottom;
- do not allow anyone to work on a ladder without a handhold;
- check the state of ladders regularly, to ensure they are in good condition;
- train people to set up ladders at the correct angle: ladders should have a slope of four units up to one out from the base;
- train people to avoid setting ladders against fragile surfaces, or on insecure footings;
- make sure the ladders used are right for the job.

Self-assessment 4 on page 82

1 The immediate actions are:
 a Assess the situation.
 b Make the area safe.
 c Give first aid.
 d Get help.

2 Investigators need to find out:

- the sequence of events that lead up to the accident;
- the system of work that was in operation;
- about any relevant environmental factors;
- what plant and equipment was involved;
- who was present.

They should also:

- take photographs, make sketches, and take measurements of the scene and the relevant features;
- obtain statements, as soon as possible, from all persons who were involved in, or who observed, the accident;
- review the facts, in the light of what has been learned, taking steps to resolve any inconsistencies or conflicting evidence;
- get expert help, if necessary;
- come to a conclusion, if possible, regarding the causes of the accident;
- generate a written report of the accident, which describes what happened, sets out the causes, and recommends changes to prevent a recurrence.

Reflect and review

5 Answers to the quick quiz

Answer 1　By looking into 'near miss' incidents, a lot may be learned that may help prevent an actual accident in similar circumstances.

Answer 2　Organizations need to have systems – policies, plans and procedures – that are designed to ensure that accidents do not happen. If these systems are inadequate or out of control, then accidents will and do occur.

Answer 3　When safety objectives are set, it is important to know whether, and to what extent, they are being met. To evaluate objectives, therefore, they should be compared with a quantifiable standard.

Answer 4　It means that many of the costs of accidents to an organization are not recognized, and not insured against.

Answer 5　Organizations should control the situation by taking the initiative, rather than simply reacting to what happens.

Answer 6　Acts of Parliament (statute law); common law or case law; EU Directives.

Answer 7　All employees have duties:

- to take reasonable care for the safety of themselves and others;
- to co-operate with their employer in matters of safety;
- not to interfere with or misuse anything provided for their safety.

Answer 8　A formal definition is: 'an identification of the hazards present in an undertaking and an estimate of the extent of the risks involved, taking into account whatever precautions are already being taken'.

Answer 9　You might have mentioned:

- checking that the ladder is secured;
- not working on the ladder without a handhold;
- checking to see that the ladder is not weak or damaged;
- placing the ladder at a stable angle;
- not placing the ladder against a fragile surface;
- placing the ladder on a secure footing;
- using the correct ladder for the job.

Answer 10　While static electricity isn't always dangerous in itself:

- the shock from static electricity can cause an involuntary movement which could result in an accident;
- sparks generated by static electricity can be very dangerous near flammable liquids, or organic powders.

Reflect and review

Answer 11   Suitable questions include:

- Is it really necessary to move the load?
- Does it have to be moved manually?
- What is the risk of injury?
- How can I eliminate or reduce this risk?

Answer 12   You could have mentioned: flying particles; dust; chemical splashing; flying molten metal; vapours and gases; radiation.

Answer 13   The simplest reason is that untidy workplaces may contain hazards of many kinds, including tripping, slipping, and falling hazards.

Answer 14   A good piece of advice is: 'Assume that what might go wrong will go wrong'. In other words, think of all the things that **could** occur.

Answer 15   Assess the situation; make the area safe; get help; give first aid.

6 Certificate

Completion of this certificate by an authorized person shows that you have worked through all the parts of this workbook and satisfactorily completed the assessments. The certificate provides a record of what you have done that may be used for exemptions or as evidence of prior learning against other nationally certificated qualifications.

Pergamon Open Learning and NEBS Management are always keen to refine and improve their products. One of the key sources of information to help this process are people who have just used the product. If you have any information or views, good or bad, please pass these on.

NEBS MANAGEMENT DEVELOPMENT
SUPER SERIES
THIRD EDITION

Preventing Accidents

....

has satisfactorily completed this workbook

Name of signatory   ....

Position   ....

Signature   ....

Date   ....

Official stamp

SUPER SERIES

SUPER SERIES 3

0-7506-3362-X   Full Set of Workbooks, User Guide and Support Guide

A. Managing Activities

ISBN	Title
0-7506-3295-X	1. Planning and Controlling Work
0-7506-3296-8	2. Understanding Quality
0-7506-3297-6	3. Achieving Quality
0-7506-3298-4	4. Caring for the Customer
0-7506-3299-2	5. Marketing and Selling
0-7506-3300-X	6. Managing a Safe Environment
0-7506-3301-8	7. Managing Lawfully - Safety, Health and Environment
0-7506-37064	8. Preventing Accidents
0-7506-3302-6	9. Leading Change

B. Managing Resources

ISBN	Title
0-7506-3303-4	1. Controlling Physical Resources
0-7506-3304-2	2. Improving Efficiency
0-7506-3305-0	3. Understanding Finance
0-7506-3306-9	4. Working with Budgets
0-7506-3307-7	5. Controlling Costs
0-7506-3308-5	6. Making a Financial Case

C. Managing People

ISBN	Title
0-7506-3309-3	1. How Organisations Work
0-7506-3310-7	2. Managing with Authority
0-7506-3311-5	3. Leading Your Team
0-7506-3312-3	4. Delegating Effectively
0-7506-3313-1	5. Working in Teams
0-7506-3314-X	6. Motivating People
0-7506-3315-8	7. Securing the Right People
0-7506-3316-6	8. Appraising Performance
0-7506-3317-4	9. Planning Training and Development
0-75063318-2	10. Delivering Training
0-7506-3320-4	11. Managing Lawfully - People and Employment
0-7506-3321-2	12. Commitment to Equality
0-7506-3322-0	13. Becoming More Effective
0-7506-3323-9	14. Managing Tough Times
0-7506-3324-7	15. Managing Time

D. Managing Information

ISBN	Title
0-7506-3325-5	1. Collecting Information
0-7506-3326-3	2. Storing and Retrieving Information
0-7506-3327-1	3. Information in Management
0-7506-3328-X	4. Communication in Management
0-7506-3329-8	5. Listening and Speaking
0-7506-3330-1	6. Communicating in Groups
0-7506-3331-X	7. Writing Effectively
0-7506-3332-8	8. Project and Report Writing
0-7506-3333-6	9. Making and Taking Decisions
0-7506-3334-4	10. Solving Problems

SUPER SERIES 3 USER GUIDE + SUPPORT GUIDE

ISBN	Title
0-7506-37056	1. User Guide
0-7506-37048	2. Support Guide

SUPER SERIES 3 CASSETTE TITLES

ISBN	Title
0-7506-3707-2	1. Complete Cassette Pack
0-7506-3711-0	2. Reaching Decisions
0-7506-3712-9	3. Managing the Bottom Line
0-7506-3710-2	4. Customers Count
0-7506-3709-9	5. Being the Best
0-7506-3708-0	6. Working Together

To Order - phone us direct for prices and availability details
(please quote ISBNs when ordering)
College orders: 01865 314333 • Account holders: 01865 314301
Individual purchases: 01865 314627 (please have credit card details ready)